AMERICAN EAGLE

To David
"Don't Give Up The Ship!"

Much thanks, + all my best

4/3/14

John Haley Bellamy: tintype, circa 1876.
Courtesy William A. Farnsworth Library and
Art Museum, Rockland, ME.

American Eagle

The Bold Art & Brash Life of John Haley Bellamy

James A. Craig

Portsmouth Marine Society
Publication Number 34
2014

Printed in Canada

Portsmouth Marine Society is a publication series of

Portsmouth Historical Society

PO Box 728, Portsmouth, New Hampshire 03802

www.portsmouthhistory.org

FIRST EDITION

Design by Grace Peirce (nhmuse.com)

Library of Congress Cataloging-in-Publication Data
Craig, James A.
American eagle : the brash life and bold art of John Haley Bellamy / by James A. Craig. -- First edition.
pages cm. -- (Publication ; number 34)
Includes bibliographical references and index.
ISBN 978-0-915819-42-3 (alk. paper)
1. Bellamy, John Haley, 1836-1914. 2. Wood-carvers--United States--Biography. I. Title.
NK9798.B36C73 2014
730.92--dc23
[B]
2014003826

Front cover: John Haley Bellamy, Pilothouse eagle for the tugboat *Mitchell Davis*; gilded pine, ca. 1890. [32 x 40 in.] Private collection. Photograph, David Bohl.

Back cover: John Haley Bellamy, “God is our Refuge and Strength” eagle plaque; painted and gilded pine, ca 1872–1900. Private collection. Photograph, Corey O’Neil.

Formal permissions to publish were received from:

Boston National Park Service, Charlestown Navy Yard
Boston Public Library
Historic New England
Kittery Historical and Naval Society
Maine Maritime Museum
The Mariner’s Museum, Newport News, Virginia
Mystic Seaport Museum
National Gallery of Art, Washington, D.C.
Naval History and Heritage Command, Washington, DC
New Hampton School
Old Sturbridge Village
Peabody Essex Museum, Salem, Massachusetts
Portsmouth Athenaeum
Portsmouth Historical Society
Portsmouth Marine Society
The Sawtelle Collection
Scholl Antiques
Scottish Rite Masonic Museum and Library, Lexington, Massachusetts
Shelburne Museum, Shelburne, Vermont
Milne Special Collections and Archives, Dimond Library, University of New Hampshire
William A. Farnsworth Library and Art Museum

This book is humbly dedicated to the memory of

Joseph William Pepperrell Frost
(1923–2008)

antiquarian, historian, and rediscoverer of the life and legacy of

John Haley Bellamy

and

To my father John Edward Craig, a great and true gentleman of the "Old School," who back in my childhood years first kindled within me a love for history, eloquence, art, and the written word.

Subscribers

Benefactors

Pete and Ginny Nicholas
Northeast Auctions by Ronald Bourgeault

Patrons

Dr. Geoffrey Clark and Sen. Martha Fuller Clark
The Labrie Family
Richard Candee & Robert Chase

Sponsors

Dr. Robert J. Allister
The Rev. John Miles Evans
Tom McCarron
Jane Porter
Lee and Rodney Roberts
John Stobart
Peter and Sue Vocca
Sumner and Helen Winebaum

Contents

	Abbreviations	viii
	Acknowledgments	ix
	Foreword	x
	Introduction	1
	The Lancaster Eagle 1	4
CHAPTER 1:	A Woodcarver is Born	7
	The Lancaster Eagle 2	14
CHAPTER 2:	An Emergent Talent	16
	The Lancaster Eagle 3	28
CHAPTER 3:	"Missing Years" and the American Civil War	30
	The Lancaster Eagle 4	42
CHAPTER 4:	The Postwar Years	45
	The Lancaster Eagle 5	62
CHAPTER 5:	The Return to Kittery Point	67
	The Lancaster Eagle 6	76
CHAPTER 6:	Glory and Majesty Captured in Whittled Pine, Part I: The Bellamy Eagle	79
	The Lancaster Eagle 7	110
CHAPTER 7:	Glory and Majesty Captured in Whittled Pine, Part II: Inspired Imitators, Atypical Artwork, and Other Carvings	113
	The Lancaster Eagle 8	146
CHAPTER 8:	"John's Fancy Friends"	148
	The Lancaster Eagle 9	162
CHAPTER 9:	A Life in Decline	165
	Epilogue	175
	Endnotes	187
	Bibliography	203
	Appendix	207
	Index	208
	About the Author	212
	Portsmouth Marine Society Titles	213

Abbreviations

Images Courtesy of:

- AKA Allan Katz Americana, Woodbridge, CT
- AS Arthur Swanson, Cider Hill Woodworks, Inc., York, ME
- D&C Deschenes & Cooper Architectural Millwork, LLC, Pawcatuck, CT
- DK David Kaselauskas, Kittery Point, ME
- DP D. Pratt, Framer, Kittery, ME
- HGAS Hyland Granby Antiques; Hyannis Port, MA
- HNE Historic New England, Boston, MA
- KHNS Kittery Historical and Naval Society, Kittery, ME
- MD Michael Dow, York, ME
- MHS Maryland Historical Society, Baltimore, MD
- MM The Mariner's Museum, Newport News, VA
- MMM Maine Maritime Museum, Bath, ME
- MSM Mystic Seaport Museum, Mystic, CT
- NEA Northeast Auctions, Portsmouth NH
- NGA Index of American Design, National Gallery of Art, Washington, DC
- NHHC Naval History and Heritage Command, Washington, DC
- SRMM Scottish Rite Masonic Museum and Library, Lexington, MA
- NPS National Park Service, Boston, MA
- OH Olde Hope Antiques, Inc., Solebury, PA
- OSV Old Sturbridge Village, Sturbridge, MA
- PA Portsmouth Athenaeum, Portsmouth, NH
- PEM Peabody Essex Museum, Salem, MA
- PHS Portsmouth Historical Society
- PMS Portsmouth Marine Soc., Sawtelle Collection
- PPL Portsmouth Public Library
- RSC Robert S. Chase, York, ME
- SM Shelburne Museum, Shelburne, Vermont
- UNH Special Collections, University of New Hampshire, Durham, NH
- WAF William A. Farnsworth Library and Art Museum, Rockland, ME

Photographers:

- DB David Bohl
- GA Gavin Ashworth
- KC Kathleen Craig
- RM Ralph Morang
- CO Corey O'Neil
- JP Jean Petty

Acknowledgments

Telling the tale of a man's life one hundred years after his passing is no mean feat; inevitably an author turns to others for guidance and assistance. Consequently, over the past six years I have amassed quite a list of remarkable persons to thank.

First and foremost, I wish to thank Mr. Pete Nicholas, whose boundless vision and generosity have made this book a reality.

I would also like to extend heartfelt thanks to the following individuals for their often priceless assistance in the research, writing, and publication of this book:

David Kaselauskas, Kittery Point, Maine
Richard M. Candee, professor emeritus, American and New England Studies, Boston University
Maryellen Burke, executive director, Portsmouth Historical Society
Arthur Swanson, Cider Hill Woodworks, Inc., York, Maine
Michael A. Dow, Architectural Woodcarving, LLC, York, Maine
Vincent Harrild, and Patrick Capozzi, Chelsea Clock, Chelsea, Massachusetts
Alan Katz, Alan Katz Americana, Woodbridge, Connecticut
Ronald Bourgeault and the staff of Northeast Auctions, Portsmouth, New Hampshire
Alan Granby, Hyland Granby Antiques, Hyannis, Massachusetts
Steve and Becky Blanchard, DR Fine Arts, North Reading, Massachusetts
Robert S. Chase, York, Maine
Brian Cooper, Deschenes & Cooper Architectural Millwork, LLC, and Early New England Restorations, LLC, of Pawcatuck, Connecticut
Angela Waldron, registrar, and Lorraine Delaney, assistant registrar, William A. Farnsworth Library and Art Museum
Wayne Manson, curator, Kittery Historical and Naval Museum
Dan Finamore, curator and George Schwartz, assistant curator, Russell W. Knight Department of Maritime Art and History, Peabody Essex Museum
Sandra Rux, curator, Portsmouth Historical Society
Aimee Newell, director of collections, National American Heritage Museum, Lexington, Massachusetts
Mark Nucup, Associate Curator, Mariners Museum, Newport News, Virginia
Jamie Wyeth and Mary Beth Dolan
David Bohl and Corey O'Neil, photographers
Grace Peirce, designer and production coordinator
Sue Ducharme, copyeditor
Linda Conti, proofreader
J. Dennis Robinson, SeacoastNH.com
Steve Estes, Kittery Point, Maine
Dr. R. J. Allister, Kittery Point, Maine
Dr. Benjamin Potter
William Lamprelle
Richard E. Winslow III

And, last but not least, to my wife Katie, who once again has been an endless source of strength, support, patience, and insight, never once protesting while I dragged her along on yet another of my epic writing projects. Thank you, dearest. You make this all worthwhile.

James A. Craig
Gloucester, MA
September 2013

Foreword

On the centennial of the death of John Haley Bellamy, nothing could be more appropriate for the Portsmouth Marine Society Press than to replace Yvonne Brault Smith's out-of-print *John Haley Bellamy: Carver of Eagles* (publication no. 1, 1982) with a new and improved study on this local legend.

Centennial exhibitions are a staple of the Portsmouth Historical Society and its partners. In the past decade we have celebrated both Treaties of Portsmouth (1905 and 1713) on their recent 100th anniversaries in our John Paul Jones House. The 2014 Portsmouth Historical Society retrospective exhibition of Bellamy's many iconic eagles and other woodcarvings opens in the Society's Discover Portsmouth, the renovated 1810 Portsmouth Academy building that formerly served as the city's library, 100 years and one day from his passing.

The merger of the Portsmouth Marine Society's press with the Portsmouth Historical Society, beginning with the publication of *Maritime Portsmouth: The Sawtelle Collection* (publication no. 31 in 2011) brought renewed attention to the examples of Bellamy's woodcarvings that the Sawtelles had acquired from New Castle, New Hampshire neighbor, Joe Copley. For that book James A. Craig wrote expert catalog entries for both Bellamy and his friend, George S. Wasson. His new research led to an exciting lecture at Discover Portsmouth, and eventually to this book.

While others have investigated aspects of Bellamy's life, Jim Craig has pulled all the available documentary and material evidence together to retell his brash life in full detail and brought a curator's eye to the bold art of this pioneer icon maker. This new study includes examples of his work that were not known thirty years ago, and makes extensive use of two collections of written documents that had not yet found their way into research libraries or museums.

Threaded throughout this handsome volume, too, is the long and complicated story of John Haley Bellamy's largest and most famous work, the *Lancaster* eagle. This figurehead, which now graces the entrance into the Mariners' Museum at Newport News, Virginia, must count as both his largest and most impressive sculpture.

Moreover, Jim Craig follows Bellamy's influence upon other woodcarvers; not only his contemporaries like Wasson, Ivah W. Spinney, and Capt. Edward Adams, but contemporary artists like Artie Swanson, Michael Dow and Dave Kaselauskas in southern Maine, and others from 'away' who have offered the American public his most famous patriotic symbols, the "Portsmouth" and "Kittery" eagles, for nearly the whole century past. These are just the kind of Portsmouth stories we exist to tell!

Richard M. Candee
President, Portsmouth Historical Society

Introduction

Gilded wings spread wide in regal majesty . . . A razor-sharp beak emitting a fierce scream, defiant and proud . . . The flag and shield of the American nation held safely in the steely grip of a raptor's talons . . . Such are the trademarks of one of the most recognized and sought-after pieces of American art: the Bellamy Eagle. Widely considered to be the pinnacle of American woodcarving, the Bellamy Eagle vies with such hallowed icons as the Statue of Liberty, the Great Seal of the United States, and Old Glory itself in representing the United States. Yet what of its creator? Who was the man responsible for giving flight to this most celebrated piece of Americana?

John Haley Bellamy was the son of a housewright, boat builder, and inspector of timber who grew up in the very bosom of a nineteenth-century shipbuilding community. It seems almost preordained that he should have turned to woodworking for his livelihood. Seldom was there a day when this quixotic man from New England did not awaken to the sound of hammers and saws transforming raw timber into oceangoing vessels. His was a world of wood, a time and place where almost every man learned to work with his hands from a young age. It was also a time of astounding advances in science and technology, of ardent patriotism and boundless optimism. His era witnessed the United States rise from the ashes and madness of a civil war to claim its place as a preeminent global power. With such a potent mix of circumstances at hand, the stage was set for the art of John Haley Bellamy to flourish.

Bellamy's artwork was fresh, innovative, and, by his own words, "original with me and never known or heard of until I produced it."[1] Yet the story of John Haley Bellamy is far more than that of a simple woodcarver. A prolific artist by his very nature, he strayed well beyond his iconic eagles to craft and design furniture, clock cases, carvings of a whimsical nature. He also patented a horde of inventions. Proving the age-old adage "No man is an island," Bellamy enjoyed a number of meaningful friendships, associating with several of America's leading late-nineteenth-century literati, including William Dean Howells, Mark Twain, and Winslow Homer.

Less than five decades after his passing, carvings by Bellamy that cost one dollar when first crafted began to sell for as much as one hundred dollars at auction.[2] Today those same works can easily fetch as much as $160,000, while larger and more exquisite pieces have sold for more than $660,000.[3] Yet while their value seemingly increases with each passing year, our understanding of this great artist has only now begun to truly blossom. Depicted alternately as a disciplined laborer and a helpless drunkard, a manic inventor and an aloof poet, an irresponsible pleasure-seeker and a devoted kinsman, his true life's story has lain obscured behind a smokescreen of ribald legends and tall tales that arose during his later years and flourished with his passing.

In exploring the life of John Haley Bellamy, two primary sources answer the questions and dispel the myths that swirl around him. One of the sources is the reminiscences of his nephew, Moses Victor Safford. Safford's writings on Bellamy have presented scholars with a unique challenge. With his personal connection to Bellamy and ease of access to other family members, Safford stands as the first and foremost independent commentator on the intimate details of John Haley Bellamy's life and personality, providing candid, comprehensive facts without which his story could never be told. Yet while Safford is the primary source for these insights, certain key elements of his recollections simply do not stand up to scrutiny and recent findings. As well, by his own admission Safford felt he was not the ideal candidate to tell his uncle's tale.

> The role of a biographer of John Haley Bellamy has been put up to me because I might reasonably be expected to tell more about him that anybody else now available. My limitations for the part will appear from the following . . . During my boyhood when going to school in Kittery and in Portsmouth I saw Bellamy at least once a week at work doing woodcarving in the Navy Yard or in his own shop. From 1886 to 1890 I saw him several times a year. From 1890 until his death the intervals between my visits were much longer but whenever in Kittery I called on him as everybody else did who had ever known him did. Such contact with Bellamy hardly qualifies me to be his Boswell and my dependence upon older contemporaries and other members of our family is increased because when I had opportunity to know him best I doubtless was influenced by the attitude of the family toward him. I never imagined that anybody would want to know anything about John Bellamy some twenty five years after he was dead.[4]

(Safford, Moses Victor: *John Haley Bellamy, The Woodcarver*. Unpublished manuscript, dated November 23, 1941, collection of the Mariner's Museum, Newport News, VA, Pg. 3, hereafter cited as MVS: *JHB*.)

As we shall see, despite his misgivings and the presence of discrepancies in his words, Moses Victor Safford's testimonies remain a cornerstone in constructing our understanding of his uncle, the great woodcarver.

The other starting place for launching an investigation into Bellamy resides within the personal writings of John Haley Bellamy himself. Beginning in his years as a student and continuing until the last night of his life, Bellamy would pen a host of letters, doodles, and poems that serve as windows into his everyday comings and goings. Ironically, for many long years these priceless documents were believed to be lost. As Moses Victor Safford told it in *The Eagle Figurehead of the USS Lancaster* (Unpublished manuscript, July 18, 1938, collection of the Mariner's Museum, Newport News, VA, Pg. 7. Hereafter cited as MVS: *Eagle*.) Bellamy's "personal records and most cherished possessions" were sold to a junkman soon after his passing in 1914. Another version of this tale claimed that another of Bellamy's nephews, Eugene Williams, "sold all of the belongings of John H Bellamy wood carver that he could dispose of . . ."[5] Mercifully; both legends proved erroneous with the discovery of Bellamy's personal papers in 1944 by Joseph Frost, the capable antiquarian of Bellamy's home town. This fortuitous discovery provides a glimpse into the inner workings of Bellamy's mind, betraying just how he viewed his maritime world and his place within it.

Today these precious clues into Bellamy's life and career are beyond the reach of junkmen and family members looking to pawn them off, kept safely within the collections of three major cultural institutions. The vast bulk of Bellamy's letters, photographs, and mementos can be found in the archives of the William A. Farnsworth Library and Art Museum in Rockland, Maine. Another cache of letters (mainly concerning his patents, inventions, and business agreements) can be found at the Peabody Essex Museum in Salem, Massachusetts, while the Dyer Library of Saco, Maine, holds an unexpected treasure trove of correspondences and manuscripts relating not only to the great artist but his celebrated father as well.[6]

Until now Bellamy's personal letters have been the least looked at of all the clues to reconstructing his life. Yvonne Brault Smith made use of the Farnsworth's collection for her 1982 book, *John Haley Bellamy, Carver of Eagles*, the first attempt at penning a comprehensive biography on Bellamy. Yet when those documents are reviewed, it is apparent that Smith skimmed their surface, neglecting to explore the collection in depth or comprehend its greater implications. Meanwhile, the holdings at the Peabody Essex Museum and Dyer Library collections have never been published until now, having only recently become accessible to scholars.

The inability of previous writers to fully draw upon these resources has left a large void in our understanding of this master woodcarver. Today, one hundred years after his passing, questions remain. Just who was John Haley Bellamy? What compelled him to devote his life to carving that prime symbol of American sovereignty, the American eagle? Should we today view Bellamy as an artisan of the highest caliber or merely as an accomplished craftsman? Should his famous eagles be classified as fine or folk art? And lastly, what is the source of Bellamy's enduring appeal?

It is my sincere hope as the author of this book that employing the latest available artifacts, cutting-edge research methods, and good, old-fashioned detective work will lift the veil of mystery surrounding America's greatest woodcarver, shattering the many persistent myths that haunt our image of him, and lead to a greater appreciation of this truly versatile man and his art.

James A. Craig

The Lancaster Eagle Part 1

On a hazy, humid July day in 1875, the warship USS *Lancaster* arrived in the lower harbor of Portsmouth, New Hampshire.[1]

The venerable old vessel, under the command of Captain Francis A. Rowe, was towed through calm waters and still air, bound for the Portsmouth Naval Shipyard. Her ironclad hull and wooden deck, scarred and blanched from eighteen years of near-constant voyaging, mutely attested to her service. Former flagship of the US Navy's Pacific Squadron, she had been a regular sight cruising the waters off South and Central America, Mexico, and California, performing a wide array of missions. Some of her more notable successes included staring down hostile Indians determined to burn the Mexican port of Alcapulco to the ground, and capturing a party of Confederate officers bent upon seizing Union gold shipments sailing from California outside the Bay of Panama in 1864. From Rio de Janeiro she had commanded the Navy's South Atlantic Squadron for half a decade, and most recently her mere presence down in the Florida Keys had helped to avert a war with Spain.[2]

Yet as the USS *Lancaster* pulled into Portsmouth harbor on that July day, she was heading toward an uncertain future. Despite her many accomplishments and proud record of service, it was decided that the *Lancaster*, like so many of her sisters, was to be decommissioned and placed "in ordinary," a limbo-like existence where she would sit, unused and indefinitely ignored. These were troubled times for the United States Navy. A crippling economic depression, budget cuts, a hostile Congress, and an apathetic public had combined to leave the Navy desperately strapped for cash. Pay was meager for officers and sailors alike, morale was almost nonexistent, ships were falling apart from age and fatigue, and all while the Committee on Naval Affairs called for a host of base closures, demobilizations, and abolishments that, if enacted, could spell the very end of the United States Navy itself.[3]

After she rode at anchor in quarantine for several days, the process of decommissioning began in earnest with the removal of her topmast, running rigging, and spars. By November 22 of that year, the once-mighty *Lancaster* found herself being towed ignominiously to her berth, where the ghoulish process continued. Soon cannons, munitions, and nearly anything else of value and utility were removed. Decks were salted against dry rot, seams caulked, and every inch of exterior woodwork slathered with varnish and tar.[4] Her name stricken from the Navy register, she was now little more than a floating hulk, bobbing at the whims of the tide, deprived of a purpose and stripped of her identity.

However, at this, her darkest hour, Fortune smiled upon the *Lancaster*. Denied the funds to build new warships by a heavy-handed Congress, the American admiralty had struck upon a novel solution: instead of replacing their outdated warships with new vessels, the Navy would instead completely rebuild those vessels already in existence from the bottom up. Under the surreptitious guise of repair work, the *Lancaster* was one of several ships to be "skinned down to their timbers" and fully reconstructed from the keel all the way to the topmast, in the process becoming a virtually new vessel (MVS: *Eagle*· Pg. 15). By December 1876, detailed plans for rebuilding the former flagship had been submitted to Isaiah Hanscom, the Navy's chief constructor in Washington, D.C.[5]

Whether conceived in celebration of her new lease on life or simply as an excuse to put as many men to work as possible in those times of economic dearth, the resolution to fix a figurehead upon the bow of the *Lancaster* had been made. It was an unusual decision, for the warship had never sported such a device before. The ostentatious ornamentation of naval craft—both military and civilian—was a practice increasingly on the wane over the past few decades. Indeed, the choice to place a figurehead on a United States warship could well be considered a quaint throwback to earlier days in the face of looming changes. Regardless of the reasoning, it was obvious that the job of fashioning and securing the figurehead could not fall upon the shoulders of a mere amateur. The man who would create this figurehead would have to be thoroughly versed in a vanishing craft. Like figureheads themselves, he would be a throwback of sorts, a man who worked with his hands in an age more and more dominated by machines, one who fashioned wood in a world of iron. He would have to possess knowledge of practical engineering combined with a profound sense of aesthetics. Of all the men then employed at the Portsmouth Navy Yard, only one could fulfill this list of requirements. Only one man could be entrusted to perform this difficult task and complete it to everyone's satisfaction—a man by the name of John Haley Bellamy.

Fig. 1.1 Gleason's Pictorial, *Portsmouth, New Hampshire, Viewed from the North*: lithograph, July 23, 1853. *DP.*

CHAPTER 1

A Woodcarver is Born

THE STORY OF JOHN HALEY BELLAMY BEGINS on Saturday, April 16, 1836, in the tiny coastal neighborhood of Kittery Point, Maine.[1] An oceanside village situated at Maine's southernmost point within the larger town of Kittery, directly across the Piscataqua River from Portsmouth, New Hampshire [Fig. 1.1] and a stone's throw from the Portsmouth Navy Yard, Bellamy's hometown was to have a decisive impact upon him, gently yet thoroughly honing him into the great woodcarver we recognize today. Indeed, to fully understand the life and art of John Haley Bellamy, it is imperative to first recognize the influence Kittery Point and its distinctly maritime culture had upon him, for without Kittery Point there truly never would have *been* a John Haley Bellamy.

Bellamy's hometown has long enjoyed a deep relationship with seafaring, beginning in the year 1603 when the English explorer Martin Pring first sailed into the area on a quest to obtain medicinal sassafras.[2] Other explorers soon followed in Pring's wake, including Captain John Smith, who in 1614 voyaged northward from Virginia into New England waters on a quest "to take whales, and make trials of a mine of gold and copper."[3] Though he found no precious metals and the whales he pursued proved elusive, Smith discovered the waters off Kittery to be teeming with something just as valuable: fish. Soon a steady stream of British colonists began to pour into the area, determined to fish the abundant nearby waters, as well as harvest sassafras and lumber, all of which they would ship back to England for a handsome profit. In time, however, the people of Kittery no longer contented themselves with merely sending raw lumber back to England for British shipyards to use as they pleased. Opting instead to establish their own local yards, Colonial-era Kittery rapidly became a premier locale for the building of commercial vessels for New England's burgeoning merchant class and mighty warships for the Royal Navy (MVS: *JHB*, Pg. 9–10).

Fig. 1.2 The Kittery Point waterfront as it appears today, as viewed from John Haley Bellamy's bedroom window. *KC (2008)*

The seamen, petty officers, shipwrights, and artisans who settled the Kittery area created a community that revolved heavily around the sea. Fishing, shipbuilding, and the lumber industry were still the epicenter of daily life in Kittery when John Haley Bellamy arrived upon the scene that Saturday in April 1836. Consequently, the sea was to have an early and profound effect upon his development. From everyday words like *krawm* and *sculch* to phrases such as "plauggy-hubbly," "chowy-hubbly," "black dreapin' fog," "nail sick," "thick as mashed mud," and "he's slipped his mooring,"[4] the young Bellamy was exposed to the colorful idioms and ideals unique to the mariner's livelihood. With a major US naval installation right down the road, throughout his life Bellamy "was probably never more than a few months in all out of the sight and smell of a man-of-war or out of the sound of a bugle" (MVS: *JHB*, Pg. 11). Many of his neighbors had once served aboard those men-of-war, fighting from the decks of such storied ships as the USS *Ranger* and *Bonhomme Richard* under the command of John Paul Jones. The future woodcarver regularly heard the adventure-filled tales of Kittery Point's mariners while growing up.

"In John Bellamy's boyhood they were still talking in Kittery about Admiral Vernon's expedition to Cartagena in 1741, in which Kittery men took part," his nephew remembered, adding:

> At the home of a boyhood friend where young Bellamy often went a sword hung over the mantel. It was the sword of an officer who had died with Nelson at Trafalgar. John and his friend would often listen as the officer's nephew, an old man then, would tell the story of this famous fight. Jefferson's embargo drove Bellamy's townspeople into privateering and thus Bellamy came to know more about Dartmoor Prison then his county jail.[5]

Kittery Point's dual culture of seafaring and shipbuilding was a romantic, all-encompassing affair. For the future woodcarver, this maritime world was inescapable, for even when he was within his family home he could not help but behold the pageantry of the village's nautical ways. Perched only a few hundred feet from the hamlet's busy Sir William Pepperrell/West Indies Wharf, Bellamy could experience the sights, sounds, and smells of the Kittery Point waterfront just by walking out his back door or opening his bedroom window [Fig. 1.2]. From coastal traders loading and unloading their mundane cargoes to fishing schooners laden with cod and vessels just back from the Caribbean loaded to the gunnels with sugar, molasses, spices, and exotic fruits, the young Bellamy was exposed to the world of the sailor in all its intimate details. His playground was the wharf, his playmates the sons of sailors, ship owners, and shipbuilders (MVS: *JHB,* Pg. 11). Yet lest we think of John Haley Bellamy's childhood neighborhood as being a dreamy, halcyon place, the words of Mark Fernald, a previous tenant of the same house Bellamy lived in, inform us that he was growing up in a community composed of genuine people, complete with quirks and all-too-human flaws:

> . . . the ancient public seat of Sir William placed me in the way of almost everything vain and evil. The place,

Fig. 1.3 The Pepperrell Mansion, ancestral seat of Sir William Pepperrell and home to the Bellamy family, as it appeared in the late nineteenth century, when Bellamy was an active resident. *PA.*

was a resort of all classes of men and boys, being at the "Harbor" (Pepperrell's cove) and the next house was a tavern for dancing, drunkenness and lewd sea-faring men.[6]

John Haley Bellamy's childhood home, "the ancient public seat of Sir William," [Fig. 1.3] further immersed the future woodcarver within the world of artistic carving. A storied edifice well before he was born, in the days before the American Revolution the house was ancestral home of Kittery's Sir William Pepperrell, a powerful merchant and commander of the Colonial forces that famously captured the French fortress of Louisburg in 1745. Originally built about 1700,[7] for many decades the mansion "surpassed in grandeur any residence in the province," given its liveried servants, ornate gardens, deer park, and, perhaps most tellingly, its "magnificent interiors with woodwork that has never been excelled."[8] Repeatedly celebrated over the centuries for its "heavy panels and quaint carvings," we find that architects as late as 1903 were still admiring and even imitating its hall and main staircase.[9] Consequently, for a young boy like Bellamy growing up within a structure replete with fine eighteenth-century raised paneling, it was in many ways inevitable that the Pepperrell Mansion should exert a powerful influence over the future woodcarver's sensibilities.

Perhaps it was only fitting that the man who would one day fashion America's finest figurehead should have lived in the Pepperrell Mansion, given that Sir William Pepperrell was one of the first merchants to ever commission a figurehead in the American colonies.[10] Yet beyond his growing up in the bosom of a shipbuilding community and in a house known for its extraordinary woodwork, one other factor that stands as a profound, if not decisive, influence for steering John Haley Bellamy toward a woodcarving career was his family.

Fig. 1.4 & 1.5 Left: The Honorable Charles Gerrish Bellamy, father of John Haley Bellamy, in his middle years: daguerreotype, circa 1860. Right: Frances Keene Bellamy, mother of John Haley Bellamy, approaching her later years: daguerreotype, circa 1860(?). *WAF*.

The union of Bellamy's father, Charles Gerrish Bellamy [Fig. 1.4], and his mother, Frances Keene Bellamy [Fig. 1.5], was a merger of two dynamic familial legacies. It was also a union initially born out of tragedy. On September 22, 1831, Frances Keene Bellamy's first husband, a house and ship carpenter and master mariner by trade, died suddenly.[11] It is uncertain what first drew Frances Keene Bellamy and Charles Gerrish Bellamy together, although given the fact that her first husband had been *his* brother John, it seems likely that the marriage was born out of abeyance to Biblical injunction and the popular conventions of the nineteenth century. Regardless, it is obvious that genuine feelings existed between the widow and her second husband. Not only would each remain steadfastly devoted to the other for the remainder of their lives, they would also help bring into this world and raise no less than nine children, the first of which was John Haley Bellamy.[12]

JOHN HALEY BELLAMY INHERITED A FAMILIAL LINEAGE with long ties to the Kittery area, particularly through his mother, who was a direct descendant of the first settlers of the Piscataqua region (MVS: *JHB*, 9). From his father he absorbed a tradition of enterprising behavior, hard work, and shrewd business sense, traits that assured his success as a woodcarver. Hailing from a Huguenot family with roots in the rich wine-producing regions of southwestern France, Bellamy's paternal ancestors possessed a potent combination of restlessness and a single-minded determination to succeed.[13] Having originally departed France for England after the siege and surrender of La Rochelle in 1627–28, the first of the Bellamy clan made their way to the New World in the eighteenth century. Armed with a coveted Royal land grant, they numbered among the early settlers of the region west of Dover, New Hampshire, where they soon developed a reputation for being a "progressive enterprising people" (MVS: *JHB*, Pg. 6–7).

Unlike his mother's family line, John Haley Bellamy's paternal ancestors did not begin to reside in the Kittery area until the early years of the nineteenth century (MVS: *JHB*, Pg. 8). As newcomers, they were first viewed with suspicion, gaining a reputation for being not only enterprising and industrious but aggressive and insistent too (MVS: *Eagle*, Pg. 3). This penchant for forceful business dealings earned the family more than its share of envy and outright enmity, with several in the village dubbing them "pirates." As Bellamy's nephew Moses Victor Safford explained it:

> In my boyhood it was not unusual to hear the Bellamys referred to as pirates, not merely because of their dark complection [*sic*] and dictatorial manner seemed suited to such an appellation but because when the Bellamys came to Kittery from New Hampshire at the beginning of the 19th century the name was unknown in Kittery except as that of a notorious pirate, Samuel Bellamy . . . (MVS: *JHB*, Pg. 8)

Whether they were indeed piratical in their business transactions or merely successful beyond normal measure, it is obvious that the Bellamys were driven to thrive. John Haley Bellamy's father was no exception to this family trait. Described as "one of the mos [*sic*] influential men in Kittery in the old days before the Civil War,"[14] and "a Warwick in Maine politics" (MVS: *Eagle*, Pg. 4), his political career began in 1830, when he served as an officer in the First Company of the First Regiment of the First Brigade of the Maine State militia.[15] From holding a number of local offices, including that of Kittery's treasurer and tax collector,[16] justice of the peace, county sheriff, state treasurer, representative and state senator, to serving as a US marshall and coming within one vote of gaining the Democratic Party's nomination for the US Congress,[17] the ambitious Bellamy arose to become one of the most trusted of Maine's public servants. One telling example of the high esteem in which he was held occurred in 1842, when he found himself personally appointed by the state's governor to select commissioners for settling boundary disputes and the rights of American fishing vessels in Canadian waters with Great Britain.[18] Another came five years later, when he was appointed by the Maine state legislature to meet President James K. Polk's presidential party at the state line for purposes of tendering "the hospitality of the state to the President and suite," a feat "which Captain Bellamy did in a most pleasing manner."[19]

With a wealthy, politically well-connected father providing him with a solid model of ambition and industriousness, it is easy to see where John Haley Bellamy gained the drive, discipline, and determination necessary to make his own phenomenal mark in the world. Yet Charles Gerrish Bellamy imparted more than a work ethic onto his eldest son. In fact, we find him to be the principal force behind steeping John Haley Bellamy in the world of the woodworker. While Kittery may have set the stage for his entrance into the field, and though the Pepperrell Mansion may have served as a profound source of inspiration for the young Bellamy, it was his father who served as the instrument for inculcating him with a practical knowledge of decorative woodcarving, ultimately setting him off along the road that would lead to his immortality.

Fig. 1.6 John Haley Bellamy, *The Flying Dragon*: watercolor on paper, ca. 1842. *WAF*.

Before he ever launched himself into politics, Charles Gerrish Bellamy had been a prolific woodworker. A self-described "house and ship joiner by trade," he had received valuable training early in his life with Boston's master builders Kimball & Webster.[20] By 1834, even as his stature in politics began to solidify, Charles Gerrish Bellamy sought to augment his training by studying for six months under a John Curtze or Kurtze, whom he identified as having been "a celebrated architect."[21] Under the tutelage of Kurtze [later Kutts], the Bellamy family patriarch learned civil architecture and practical geometry, two subjects that would serve him well as he engaged in building houses, public buildings, boats, and bridges.

Kittery was (and still is) littered with the handiwork of Charles Gerrish Bellamy. Regarding his prolific career as a builder of houses and boats, it has been said:

> He did build houses and they still stand to prove that he was an architect of excellent ability and good taste. I never heard that he was engaged in the building of any of the many ships that were built on the Piscataqua in the last century but he did some boat building and at different times he was in charge of the boat shops at the Boston and at the Portsmouth Navy Yards. He is credited as being the originator of the "Portsmouth boats" which long had a reputation for tightness, strength, and durability (MVS: *JHB*, Pg. 8).

Other ambitious projects he conceived, designed, and oversaw involved the construction of the Kittery Point Bridge, a pile bridge connecting Kittery Point directly with the mainland in 1837, and the Boston and Maine Railroad bridge, the second bridge to ever span the Piscataqua River.[22]

With a highly skilled woodworker for a father, it was in many ways assured that John Haley Bellamy (as well as his brothers Charles Jr. and Elisha) should pursue the woodcarver's calling over that of the sailor's. In his carpenter-architect father, Bellamy found a strong, successful example of manhood to emulate, all while his family recognized, supported, and earnestly encouraged his artistic talents (MVS: *JHB*, Pg.4). However, the earliest known examples of the future woodcarver's creative output are not woodcarvings at all, nor do they involve eagles.

Painted when he was a little boy, pictures like his watercolor *The Flying Dragon* [Fig. 1.6], with its vibrant colors and mythological subject matter reveal that a strong sense of whimsy was present in Bellamy from early childhood. Furthermore, Bellamy's disciplined, competent script showcases his natural flair for penmanship, a talent that would mature as he grew older. *The Flying Dragon* teases the viewer with suggestions that its creator has been influenced by the rich tradition of marine folk art; in viewing it, the whales, sea serpents, mermaids, and other fantastical beasts that sailors etched upon whale teeth and walrus tusks, inserted into the corners of their maps, and tattooed upon their flesh for centuries readily come to mind. A boyhood fantasy on crude brown paper, it confirms suspicions that Bellamy held artistic proclivities from a young age and supports the notion that indeed his nautical neighborhood contributed an early influence.

Bellamy's early childhood education was within the Kittery district school system and at home from private tutors (MVS: *JHB*, Pg. 5). He most likely attended a two-room structure that featured multiple grades of boys and girls receiving instruction within the same class space.[23] As was typical in Kittery Point at that time, John Haley Bellamy would have started going to the district school at around the age of five or six, with his studies concluding anywhere between the ages of nine and thirteen, when it was customary for Kittery boys to embark upon their first voyage as a cabin boy aboard coastal traders.[24] It appears that Bellamy's education—both public and private—instilled within him a healthy respect and love for learning, for as his nephew recalled: "Throughout his life he was a student of history, literature and art. He was much interested in heraldry" (MVS: *JHB*, Pg. 5).

Remembered in later years by one of his childhood chums as "a manly little boy," Bellamy exuded an amiable charm, one that easily earned him many friends (MVS: *Eagle*, Pg. 5–6).

Fig. 1.7 John Haley Bellamy as a young man: ambrotype, circa 1856. *WAF.*

Additionally, he appears to have been a favorite within his large family, living "most harmoniously" with his parents and numerous siblings (MVS: *JHB*, Pg. 4). Emerging into his adolescent and early adult years, Bellamy's physicality and personality began to solidify and take shape. Though he was described as being "an expurgated pocket edition of his father," who, "like his father . . . was fond of good company, good food and of all the comforts and luxuries of civilization" (MVS: *Eagle*, Pg. 5), several telling differences between father and son began to emerge. While they may have shared the same height, Bellamy was much smaller and slenderer than his barrel-chested father. And though it became apparent that he "would not allow anyone to impose on him," and "could be imperious in his treatment of others without much provocation and anybody whose company John did not welcome was made to know it," Bellamy on the whole lacked his father's blustering, aggressive demeanor, being instead "essentially courteous, quiet mannered, and good natured in his dealings with others" (MVS: *Eagle*, Pg. 5–6).

Concerning his "dealings with others," the following priceless image of the future artist perched on the cusp of manhood has survived:

> During his early manhood he was companionable, jovial, and popular with men. The girls, too, liked him. In fact, according to one of his contemporaries, he was very handsome, and any damsel favored with his attentions regarded herself as fortunate. In the prime of his life, his countenance and bearing suggested those of an old French *chevalier* . . . Black curly hair, steady, deep-set dark eyes, a Roman nose, an alert manner, a slight swagger, and withal a heavy, reddish-brown moustache and imperial made John Bellamy a striking figure in any company.[25]

These descriptive words are confirmed by photographic evidence. Taken in his late teens/early twenties, a photograph of Bellamy [Fig. 1.7] seconds those claims that he was "inclined to be striking in his attire but within the bounds of good taste," and much admired by women (MVS: *Eagle*, Pg. 5). Radiating an aura of dark, even exotic, romance and hailing from a prominent, successful family, John Haley Bellamy seemed well poised to win a fetching bride from among Kittery's most respected families.

Yet in a now-legendary turn of events that would haunt him for the remainder of his life, Bellamy's entrance into the world of courtship was destined for unreturned overtures and bitter failure. Concerning this chapter in his life he understandably remained silent, penning no known account for posterity and leaving it to his nephew to relate the events as they happened. As Safford told it, when the ambitious Bellamy had arrived at the proper age for seeking a wife, he set about courting "an estimable, vivacious, and beautiful Kittery girl of about his own age."[26] However, he was not alone in seeking this young woman's hand in marriage, for a handsome young lawyer was at that same time also making romantic overtures to her. Placing him at a further disadvantage was the fact that Bellamy was "obliged to be away from home for considerable periods of time," during this moment in his life, a situation that forced him to conduct most of his wooing via handwritten letters and passion-drenched poems.[27]

One account of the unhappy event claims that the lawyer succeeded in winning the young beauty from Kittery; another states that none of the players in the love triangle were ever to wed. Regardless, the end result left John Haley Bellamy unwed and alone, a circumstance that would persist to his dying day. Though one may assume that "his inglorious share in the contest probably pricked John's pride more deeply than it wounded his heart," by his nephew's estimations this failed venture into matrimonial waters placed the future artist in a cheerless turmoil, one wherein "friends thought that they observed a change in his disposition. He became, it is said, less ambitious and more inclined to recklessness."[28]

Whether Bellamy's pride had merely been "pricked" or he was left to nurse a shattered heart for the rest of his days, it is apparent that the failed courtship precipitated a sudden and disquieting change in Bellamy's personality that left family and friends concerned. Lost in the tumultuous, torturous haze of unrequited love and unable to regain his life's direction, dramatic efforts were required to get the future woodcarver's stray life back on course—efforts that would take him well beyond the familiar lanes and wharves of Kittery Point and into a larger, more rewarding world.

The Lancaster Eagle Part 2

THOUGH NONE INVOLVED COULD HAVE KNOWN IT AT THAT MOMENT, THE DECISION TO ADORN THE USS *LANCASTER* WITH A FIGUREHEAD was destined to give birth to a matchless work of art, one that has come to represent the highest level of achievement American woodcarvers have ever dared aspire to. While the identity of the man who crafted this singular piece of woodcarving excellence has never been in question, the identity of the person who first suggested placing a figurehead on the *Lancaster* has been a maddening mystery. Many early commentators and writers on Bellamy logically assumed that it was he who had first proposed that the *Lancaster* should sport a figurehead, including Bellamy's nephew, Moses Victor Safford. Concerning the matter, Safford claimed in 1938, "It seemed to be regarded as one of John Bellamy's ideas and everybody expected him to make good on anything which he proposed" (MVS: *Eagle*, Pg. 12). Yet in 1941, Safford, who was a regular visitor to Bellamy's studio at the Portsmouth Naval Shipyard during the time the *Lancaster* figurehead was being constructed, changed his mind completely on the matter, choosing instead to assert, "With whom the idea originated of putting an eagle figurehead on the *Lancaster* I never heard, so far as I can now recall" (MVS: *JHB*, Pg. 13), adding:

> He [Bellamy] was never in a position to dictate the features of a naval ship, ornamental or otherwise . . . The Naval Constructor or somebody else in authority decided on the general character of the ornamental work to be put in a ship under construction or repair and if it involved wood carving Bellamy was employed at the Navy Yard and turned loose on the job . . . His personal relations with those in positions of authority were such that he was probably consulted regarding proposed ornamental features of ships or of Navy Yard buildings but with his naval background it is safe to say that he would not make suggestions without intimation that suggestions would be acceptable.[1]

It is this version of Safford's testimony, in particular his suggestion that a "Naval Constructor or somebody else in authority" was responsible for proposing the figurehead, that ultimately proves correct. As revealed within an obscure interdepartmental letter dating from 1876, the man responsible for initially conceiving the idea of the USS *Lancaster* figurehead was not Bellamy but rather a US naval constructor working at the Portsmouth Navy Yard at that time by the name of T. D. Wilson:

> Naval Constructor T. D. Wilson, of this Yard, has prepared a Model, and plan of alterations which he submits to the Bureau in rebuilding the USS "Lancaster" and I respectfully request permission to send them to the Bureau for its consideration. [2]

With the approval of T. D. Wilson's plans and his apparent choice of John H. Bellamy as the carver to execute the eagle figurehead,[3] so began the long, arduous process of design and planning. According to Bellamy's nephew, "The whole business started with the laying out of molds just as the construction of a ship starts" (MVS: *Eagle*, Pg. 12). Again Safford's words ring true, for the rule in designing figureheads in the eighteenth and nineteenth centuries called for an intensive collaborative effort between the ship constructor and ship carver, one which began with the constructor approaching the carver with the proposed figurehead design and laying out the main lines of the ship's fore-part in chalk on the carver's studio floor, all "so that there would be no misunderstanding as to the placing of the figurehead."[4]

A priceless surviving pencil sketch drawn by Bellamy on heavy brown paper mutely testifies to this process, revealing Bellamy's initial thoughts as he and Wilson discussed and plotted out the *Lancaster* Eagle figurehead in surprisingly spartan detail. In preparing and planning this figurehead, Bellamy found himself dependent upon T. D. Wilson for pertinent information concerning the *Lancaster's* bow assembly. However, once he had been given the requisite information, as per the usual custom of the time, it would have been Bellamy and Bellamy alone who was deemed responsible for installing the figurehead correctly. While Wilson may have made suggestions, Bellamy would have been left to determine its design and execution.[5]

Incredibly, despite all its involvements, its size, and its heroic subject matter, the project did not draw any particular notice at the time. Newspapers native to the Portsmouth and Kittery areas make scant mention of the *Lancaster* during her rebuilding, let alone the figurehead that was being crafted for her. It was a phenomenon that in later years was recalled with a hint of puzzlement:

> I do not remember that the construction of this huge eagle or the mounting of it on the Lancaster then being rebuilt at the yard aroused any special interest or comment (MVS: *Eagle*, Pg. 12).

With the larger world beyond the Portsmouth Navy Yard oblivious to the mighty work of art being planned on the floor of his studio, John Haley Bellamy rolled up his sleeves and set to work.

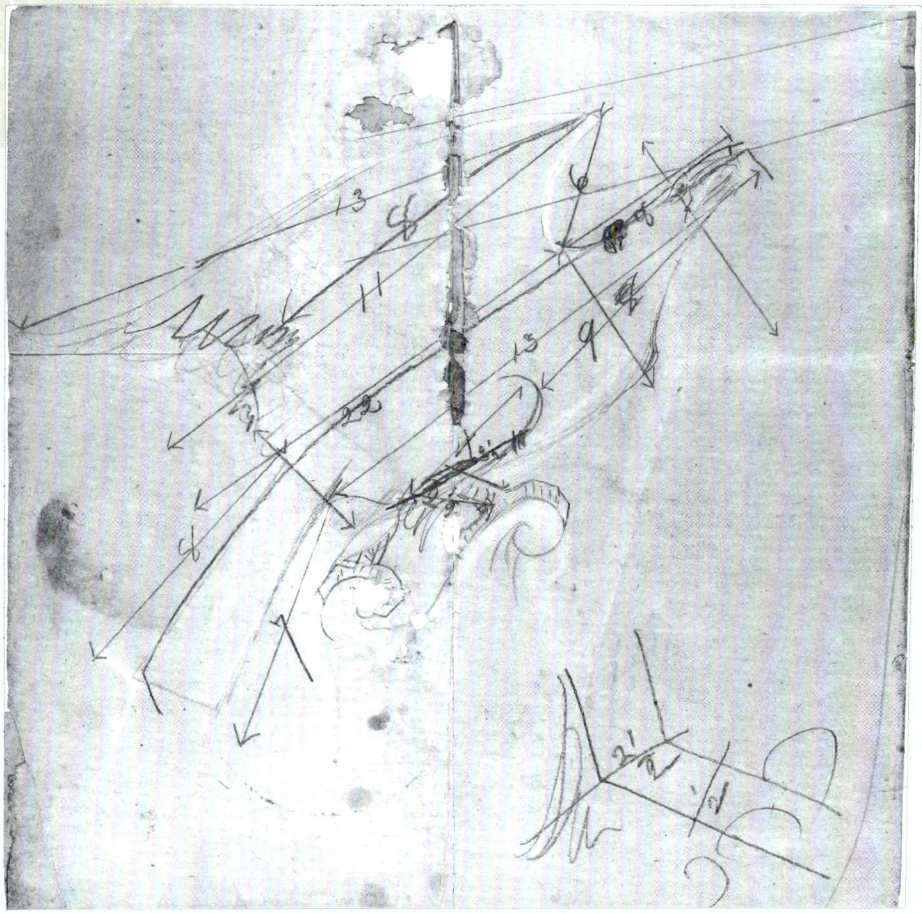

Fig. 2.1 John Haley Bellamy, *Initial schematic drawings for U.S.S.* Lancaster *figurehead*: pencil and paper, ca. 1876–77. *Private collection. MM.*

CHAPTER 2

An Emergent Talent

EVER SINCE THE NAME JOHN HALEY BELLAMY began to spill from the lips of antiques dealers and astute collectors of Americana, art historians have wondered who trained Bellamy in the ways of woodcarving. How he came to arise from the malaise and reckless abandon of his failed courtship to become one of his country's greatest artisans and wherein he derived his acute, imaginative artistic vision have never been satisfactorily answered.

In the quest to determine where the creator of the *Lancaster* Eagle figurehead gained the skills so necessary to produce this awe-inspiring piece of art, most have naturally turned to the words of Moses Victor Safford, "I was told that Bellamy studied 'art' in Boston and in New York, but was left with an impression that such courses were comparatively brief" (MVS: *JHB*, Pg. 5). Trusting in these words, many elected to unquestioningly embrace this statement as fact, echoing these words time and again with declarations such as: "John [Bellamy] . . . had a good general education and some courses in art education in New York and Boston,"[1] "As a youth he studied art in Boston and New York . . ."[2] and "Bellamy's father . . . was able to send his son off to Boston and New York to study art."[3] With each repetition, this notion has been further cemented as a Gospel truth.

However, the claim that Bellamy attended formal art academies in these cities does not stand up to serious inquiry. To begin, the very idea that he found it necessary to go to Boston and New York in order to acquire knowledge of woodcarving simply makes no sense. In nineteenth-century America, the woodcarver's trade was not an academic art form that one studied in school but instead a vocation learned on the job via the apprenticeship model.[4] To learn woodcarving, Bellamy had no need to journey far afield in search of training because there were so many opportunities to be taught the trade in his own shipbuilding hometown of Kittery, in nearby Portsmouth, New Hampshire, and within the Bellamy household itself.

Students learning sculpture in nineteenth-century American art academies were not trained in woodcarving. Rather, they were tutored in the academic sculptural traditions of Europe, where marble and granite were the preferred materials for sculptors, not wood.[5] Thus, even if Bellamy had gone off to an art school, the experience would have been all for naught, as he would have learned precious little that would have been useful in his woodcarving career. Furthermore, when Bellamy passed on in 1914, no mention was made in any of his obituaries of his having ever received such schooling. And lastly, it must be noted that it was simply impossible for John Haley Bellamy to have received formal art instruction in Boston at that time in his life, as no art schools existed in Boston during the formative years of his career.

In total, Boston was destined to host three art schools during the nineteenth century. The most famous and prominent of the three, the School of the Museum of Fine Arts, first opened its doors to students in January 1877.[6] The Normal Art School started instructing students in 1873, while the Cowles Art School began accepting students in 1882.[7] Just considering the dates when these three institutions started taking in students makes it clear that Bellamy never received schooling in that city, for they opened for business some twenty or more years after John Haley Bellamy had already begun his carving career in earnest. It is much the same concerning the assertion that he went to New York for schooling. Of the two prominent art schools operating in nineteenth-century New York, the Art Student's League of New York was founded in 1875,[8] while New York's National Academy of Design—which admittedly opened its doors in 1826[9]—holds no record of John Haley Bellamy having ever been enrolled there.

Consequently the question remains: where did Kittery's favorite son receive his training? Previous scholars have championed the notion that Bellamy learned the woodcarver's trade while under the tutelage of a particular gentleman by the name of Laban Smith Beecher.

In the year 1822, when he was just seventeen years old, Laban Smith Beecher migrated from his home in New Haven, Connecticut, to Boston, Massachusetts, in the hopes of becoming a ship carver.[10] Believed to have apprenticed under the noted woodworker Isaac Fowle, Beecher was either an exceptionally quick study or he had already received substantial training while living in New Haven. Only two years after his arrival in Boston, he was receiving prestigious commissions, most notably the honor of creating a gilded eagle to sit atop Portsmouth, New Hampshire's cherished Liberty Pole [Fig. 2.2].[11]

Dashing in appearance, well attired and appearing every inch the romantic artisan, Beecher was, however, not destined to become a great American woodcarver. His most famous creation, a figurehead of President Andrew Jackson commissioned by

Fig. 2.2 Laban Beecher, *Portsmouth, New Hampshire Liberty Pole Eagle*: gilded wood, 1824. [54 x 26 in.] *PPL / CO.*

the US Navy in 1834 to crown the bow of the famed USS *Constitution*, is noteworthy not due to its technical merits but rather because of its amusing demise. Tragically stiff, plain, and lacking in any vitality, Beecher's vision of America's seventh president would easily have been forgotten had its original head not been famously decapitated, sawed off in the dead of night by a Bostonian who despised "Old Hickory" and his politics, while the *Constitution* rode snugly at anchor in the harbor.[12]

Ultimately, Laban S. Beecher's career as a carver was destined to be short and rather infrequent, with the majority of his commissions coming from the US Navy.[13] However, his work for the navy probably first introduced him to the Bellamy family. In August 1828, Beecher was recommended by a Mr. J. D. Henley, naval officer, to the president of the Naval Board as an ideal candidate to fashion decorative woodwork for the stem and stern of the sloop-of-war USS *Concord*, which at that moment was berthed in the port of Portsmouth, New Hampshire.[14] It was not the only warship docked in that navy yard to receive Beecher's attention; Beecher also fashioned a figurehead for the USS *Santee*, and from a drawing done in his own hand it is known that Beecher at least illustrated (if not executed) a design for a figurehead to adorn the bow of the USS *Alabama*, both projects unfolding specifically while the vessels were in the Piscataqua.[15] Even the Portsmouth Navy Yard's marine barracks came to be adorned with his handiwork, eventually sporting a gilded Beecher eagle.[16]

Because Charles Gerrish Bellamy was in charge of the boat shops at both the Boston and Portsmouth navy yards, it seems quite likely that the Bellamy family patriarch first became acquainted with Beecher and/or his work via his navy yard employment, probably during the early 1830s. From this association, we can easily imagine Charles Gerrish Bellamy proposing that his oldest son work for Beecher. Yet regardless of how these two men met and whether or not his father's influence was necessary, by March 1857, when John Haley Bellamy was twenty-one years old, he was working at Beecher's Boston studio at 102 Commercial Street. That month his father penned him the following letter from the navy yard:

> Dear Son:–
> Not having heard from you by letter for some time I write you to know how matters appear at the shop of Beachers [*sic*] will he have business enough to continue your services after you are free, how does matters look there, you must come and make us a visit when your time is up whether you stop with him or not, business in all branches will be good this season, and we must all try to see how many *shiners* we can make this season, so that we can lay back next winter. I have procured a place for Charles in the Navy for this season, say nothing about this to strangers. Elisha I intend to send to school you know he is a great little fellow and I can't tell what branch of business he will excell [*sic*] in best . . .
>
> And now My Dear son write me at the receipt of this if it is but a dozen lines, tell me how matters appear at the shop.[17]

This revelation makes it apparent that John Haley Bellamy worked in Laban Beecher's shop. That the man responsible for crafting the US Navy's finest figurehead should have apprenticed in the shop of a gentleman whose carvings adorned the prow of several American warships seems only natural, especially given how the carving of figureheads "was a craft that was based on the apprenticeship system of master, journeyman and apprentice."[18] The case for Beecher having been a key influence upon Bellamy can be easily argued when one considers that he (like Bellamy) not only created ship figureheads but, perhaps most tellingly, also crafted several notable carvings of that most distinctive Bellamy motif—the American eagle. From the eagle he carved for the Portsmouth, New Hampshire Liberty Pole [Fig. 2.2] to the eagle that adorned the Portsmouth Navy Yard's marine barracks [Fig. 2.3] to an eagle that hovered over the very doorway to the Bellamy family household [Fig. 2.4], Laban Beecher showed time and again a competency and facility in rendering the American eagle in gilded pine.

Yet before crowning Beecher with the full glory of having been John Haley Bellamy's mentor, two distinct factors must be considered. First, while Beecher may have carved several eagles, there is scant stylistic similarity between his inspired but ultimately awkward birds and the forceful, masculine creations Bellamy fashioned in later years. Beecher's eagles stand as attempts at creating anatomically correct, realistic raptors, while Bellamy strived toward an altogether different art form, where cumbersome realism was discarded in favor of a more representational vision, one which conveyed to the viewer a streamlined sense of an eagle's force and presence with minimal detail. Secondly, while Bellamy was toiling away in Beecher's studio in 1857, he could not have been working there for longer than a mere six months—hardly enough time to receive a thorough training in the woodcarver's art, especially when one considers that apprentices generally trained for about five years on average before they were competent enough in the craft to work alone.[19] Moreover, Beecher himself was no longer present in Boston by 1857, having moved west and bequeathed his business to his assistant,

Fig. 2.3 Laban Beecher, *Portsmouth Navy Yard Marine Barracks Eagle*: gilded wood, ca. 1830. [52½ x 164 in.] *SM*.

Fig. 2.4 Attributed to Laban Beecher, *Pepperrell Mansion Eagle*: gilded wood, ca. 1830. [8¼ x 21¼ in] *WAF*. shown *in situ* above side entrance door to the Pepperrell Mansion, mid-twentieth century. *PA*.

a Mr. John Mason, thus rendering it impossible for him to have had anything beyond a distant, indirect role in shaping John Haley Bellamy's artistic vision at best.[20] Though certainly a man of some talent (and possibly the man who first turned Bellamy on to the American eagle as a decorative motif), it becomes readily apparent that Beecher was not responsible for schooling Bellamy in the basics of the carver's vocation.

IN TRYING TO TRULY UNDERSTAND THE EVOLUTION of John Haley Bellamy's woodworking skill, it is vital to first remember that he lived in a world of wood. In nineteenth-century America, knowing how to work wood into useful objects was an everyday necessity, just as in today's America it is essential to know how to send an e-mail, drive a car, or operate a microwave oven. From highly skilled artisans to barely literate farmers, men of the Republic fashioned a horde of everyday utilitarian objects from wood, including such diverse items as butter molds, toys, hunting decoys, and ax handles to furniture, candelabra, and weathervanes.[21] The state of Maine, with its abundant supply of timber, healthy shipbuilding industry, and large population of farmers, sailors, and lumberjacks (three classes of laborers known for turning to woodcarving both out of practical need and as a means of amusement during long periods of enforced idleness), proved to be an environment particularly amenable for encouraging woodworking activities.

In trying to identify who first introduced John Haley Bellamy to woodcarving, Charles Gerrish Bellamy, the most obvious and influential of his teachers, has long been overlooked. Bellamy's father was a thoroughly accomplished woodcarver in his own right, trained by an architect and skilled in everything from building wooden boats to wooden bridges, so his influence upon his son would have been paramount. We can easily imagine John standing beside his father as a little boy, observing him at work on a neighbor's house or building a boat in the back yard, asking questions, and, in a moment of precocious curiosity and emulation, taking up a tool and whittling out his first crude creation.

There is no reason to assume that Charles Gerrish Bellamy would not have encouraged his son's earliest endeavors, instructing, nourishing, and gently honing his talents during those crucial early years of learning. Evidence supporting this notion can be found in the fact that John and his brothers Charles and Elisha would all follow in their father's footsteps and become skilled woodworkers themselves, even toiling beside their father in their adult years on various carpentry projects. The letter that he sent his eldest son on March 25, 1857, shows that Charles Gerrish Bellamy was actively seeking training and employment within the woodworking trade for his two oldest sons and pondering what to do with his third. Upon exploring the contents of John Haley Bellamy's toolbox, we discover that it not only held tools belonging to Bellamy but to his father and brothers as well, mutely testifying to how closely the family worked together on their art and illustrating that few, if any, boundaries existed between John and his father as far as woodworking was concerned.[22]

Despite whatever exposure and impromptu schooling John Haley Bellamy received from his father over the years, the time would have inevitably arrived when he had to leave home and broaden his knowledge by entering into a formal apprenticeship with an established woodcarver. As to the identity of the man John Haley Bellamy apprenticed to, we find that sources contemporary to him make no mention of Laban S. Beecher, citing instead a versatile craftsman from neighboring Portsmouth named Samuel Dockum. As Bellamy's obituary in the *Portsmouth Times* stated on April 6, 1914, "Mr. Bellamy was a noted wood carver, learning it with the late Samuel M. Dockum . . . "[23] These words are seconded by the *Portsmouth Herald*, which on that same day issued the statement "He [Bellamy] learned the carver's trade with Samuel Dockham [*sic*] of this city and soon became an expert at wood carving . . . "[24]

Born in either Portsmouth or Greenland, New Hampshire, on January 13, 1792, Samuel Dockum [Fig. 2.5] was in

Fig. 2.5 Samuel Dockum, the talented, versatile woodcarver from Portsmouth, NH, who formally trained John Haley Bellamy during his woodworking apprenticeship: portrait, oil on board, ca. 1840. [32¼ x 27¼ in.] *PHS / CO.*

Portsmouth by 1820, when he petitioned the Pythagoras Lodge No. 33 of Portsmouth to become a Mason.[25] The next year he was listed in the Portsmouth business directories, offering his services as a cabinetmaker and carver.[26] A talented and resourceful artisan, Samuel Dockum became that city's most prosperous and visible woodcarver by the mid-1830s, owning "the largest and most prominent shop in Portsmouth, offering the widest selection of goods available in the community and catering to all levels of patronage."[27] His labors involved everything from manufacturing bureaus, Grecian couches, coffins, and rocking chairs to boot forms, worktables, ironing boards, and business signs, as well as making repairs on the organ at Portsmouth's Saint John's church.[28] He also engaged in furniture repair and sold lumber wholesale, in one instance retailing enough pine "for 100 bureaus, excepting the bottoms."[29]

Starting in 1842, the enterprising Dockum began to advertise his services as a house and ship carver, carving the occasional mantelpiece or balustrade for discriminating homeowners and executing the finish work on many of the sailing ships built in Portsmouth and vicinity at that time.[30] From his business records, it becomes evident that he produced a wide assortment of nautical ornaments, from letters, stars, and centerpieces to adorn a ship's stern to catheads, billetheads, newels, and trailboards.[31] At times Dockum's services were sought for other, more complicated commissions, such as when he finished the cabin of the clipper ship *Josephine* "in a rich but not gaudy manner, in Mr. S. M. Dockum's usual good style."[32]

Working for Samuel Dockum offered Bellamy a superb opportunity to learn the woodcarver's trade, specifically involving ship carving, at the side of an accomplished and respected artisan. Dockum was noted for hiring men and boys who brought with them a working knowledge of the craft, which reinforces the theory that Bellamy had already received significant exposure to the carver's art from his father. It is an established fact that at Dockum's shop, "employment was based on ability and not on family connections."[33]

We can surmise that Bellamy's apprenticeship lasted for a period of six years, given that surviving court records concerning other Dockum apprentices stipulate a similar tenure.[34] Payment for his services would probably have been in clothing, goods, or cash, as permitted by the usual custom for the time; working conditions generally appear to have been favorable in the shop.[35] Though occasional instances of apprentices running away exist, this appears to have been the exception rather than

Fig. 2.6 Samuel Dockum, *What-Not Shelf with Bust of a Deer*: mahogany with gilded metal, ca. 1840. [10½ x 11 x 8 in.] *PHS / CO.*

Fig. 2.7 Top: Close-up of a drawer pull on a chest of drawers carved by Samuel Dockum, plainly displaying the double parallel *S*-shaped cut that would become Bellamy's signature mark. *PHS / CO.*

Fig. 2.8 Bottom: Detail of double parallel and single *S*-shaped cuts derived from his teacher Samuel Dockum, as seen on Bellamy's *Carved Eagle with Two Draped Flags Under a Center Shield*: pine, ca. 1872–1890. *Private collection, AKA /DB.*

Fig. 2.9 Detail of double parallel *S*-shaped cuts derived from his teacher Samuel Dockum, as seen on the feathers and visage of Bellamy's *Large American Eagle Atop an Out-of-round Ball*: pine, ca. 1866–1872. *Private collection, AKA /DB.*

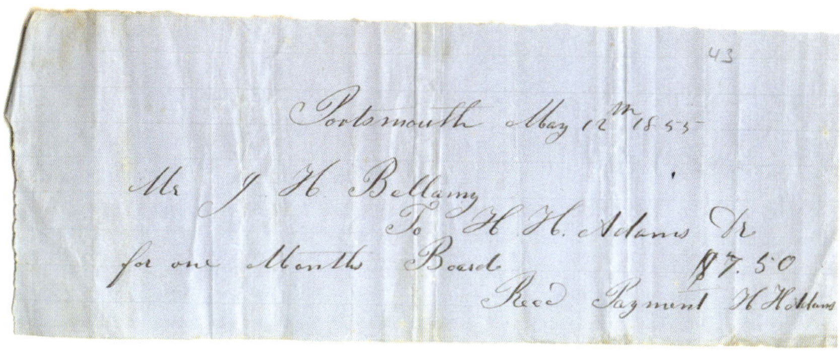

43

Portsmouth May 12th 1855

Mr J. H. Bellamy
To H. H. Adams Dr
for one Month Board $7.50
Recd Payment H H Adams

Fig. 2.10 Receipt from John Haley Bellamy to H. H. Adams of Portsmouth, New Hampshire, for one month's board, May 12, 1855. *WAF.*

the rule among Dockum's employees.[36] That Bellamy found his time under Dockum's tutelage both agreeable and of inestimable value is confirmed by Moses Victor Safford, who states concerning this period in Bellamy's life: "Suffice it to say, that the selection of both vocation and master was a happy one. Bellamy soon acquired a reputation for originality in design, skill in workmanship, and rapidity in execution."[37]

From the known surviving pieces attributed to Dockum, we gain a glimpse of the influence Portsmouth's master carver had on his young apprentice. In Dockum's *What-Not Shelf with Bust of a Deer* [Fig. 2.6] we can easily see the source from which Bellamy derived the vision for crafting the wide series of Masonic-themed and purely decorative what-not shelves he proffered to the public starting in the 1860s, as well as the origins of the practical training he needed to execute his little-known realistic animal busts of later years. In the carved drawer pulls of Dockum's shelves we discover the roots of that most telling of Bellamy's distinctive marks, the double parallel *S*-shaped cut [Fig. 2.7]. Found lurking on just about every one of Bellamy's avian sculptures, [Fig. 2.8 and 2.9] this deceptively complex incision served as Bellamy's artistic signature, a surefire (and much copied by his imitators) way for someone with a discerning eye to distinguish a Bellamy eagle carving from its lesser contemporaries. Long believed to have been unique to Bellamy, in Dockum's drawer pulls we encounter nothing less than the source from which Bellamy derived his signature mark—a testimony to how deeply Dockum's training impacted him.

Though he probably began working for Dockum in 1851, we now know from the discovery of a receipt for one month's board to a H. H. Adams that Bellamy was in Portsmouth at least as early as May 1855, when the future artist was nineteen years old, presumably toiling away as Dockum's apprentice [Fig. 2.10]. The fact that he was living and working in Portsmouth during his late teens/early twenties brings us back to Bellamy's failed courtship. Remembering that he was "obliged to be away from home for considerable periods of time," and that "he had to conduct most of his wooing by correspondence . . ." it is now clear that Bellamy was busily working for Dockum at the same time that he was seeking a bride, the reason he had to conduct so many of his sentiments via letters.

Fig. 2.11 The New Hampton School of New Hampton, New Hampshire: photograph, ca. 1854. *New Hampton School.*

the drastic measures needed to save the young woodcarver from himself.

By the mid-nineteenth century, new medical theories, improved means of transportation, and enlightened attitudes concerning illness and recuperation led to a revolution in physical and mental health care. Patients afflicted with anything from consumption to rheumatism to melancholy were increasingly advised by their doctors to travel to healing spas, mountain retreats, and other rustic enclaves on the quest "to cure or alleviate illness through a change of climate, change of scenery or active exercise."[38] In the spirit of the hour, Charles Gerrish Bellamy decided to send his son off on such a journey. The blunt, hardworking patriarch of the Bellamy clan determined that rather than send his oldest son off to restful spas and peaceful wilderness retreats, he would shock him back into life with hard work and religion.

Why his inability to capture the heart of the "estimable, vivacious, and beautiful Kittery girl" led to such a dramatic change in attitude and habit shall probably never be known. Regardless, the failed love affair rendered him a changed man. Suddenly "less ambitious and more inclined to recklessness," to concerned family and friends Bellamy seemed poised to destroy all that he had labored so hard to achieve before his career had even begun. Sensing the danger, his father implemented

This appears to have been the reason for Bellamy's brief employment at Laban Beecher's shop. Sent off to work for a trusted professional acquaintance, Bellamy's presence in Boston was designed to place some distance between him and the scene of his heartbreak while filling his days with solid labor and affording him little time to indulge in destructive behaviors or dwell on his lost love. How effective a "cure" this was we can only speculate, however. Had events turned out differently, perhaps

Bellamy would have remained employed at Beecher's for years to come, yet that was not to be.

When Charles Gerrish Bellamy asked his son to "tell me how matters appear at the shop," specifically, "will he have business enough to continue your services after you are free," he had good reason to pry, for in 1857 an economic depression struck America. The shipbuilding industry was acutely affected, having been flooded with a glut of new clipper ships at a time when the volume of trade was declining precipitously. With the rapid drop in shipbuilding (particularly clipper ships), the demand for the figureheads and rich interior appointments aboard vessels evaporated, causing more than a few trained shipcarvers to seek work in other trades.[39] Laban Beecher's shop was not immune to this trend, and by the fall of that same year we discover that John Haley Bellamy was no longer under employment there, instead enrolled in a private school up in the rural highlands of New Hampton, New Hampshire.[40]

Founded on September 17, 1821,[41] the New Hampton Literary and Biblical Institution at New Hampton, New Hampshire, [Fig. 2.11] was widely known for being "among the most favored institutions in the country"[42] for college preparatory schooling, drawing students from "every New England state, and also from New York, New Jersey, Pennsylvania, and Canada."[43] At first glance it may seem incongruous for a young woodcarver fresh from his apprenticeship and first gainful employment to suddenly be attending a "Literary and Biblical Institution." However, one can begin to understand the reasoning behind Charles Gerrish Bellamy's decision by pondering the way in which the Institute described itself in its catalog:

> This Institution is pleasantly located at New-Hampton Village, near the centre of the State. It is accessible daily from nearly all parts of New England; is six hours from Boston, via Bristol Station, Northern railroad, or Boston, Concord and Montreal Stations at Holderness and Sanbornton Bridge. The location is remarkably healthy, and free from temptations to *idleness* and dissipation.[44]

"Pleasantly located" in a "remarkably healthy" corner of New Hampshire, at first glance the New Hampton Literary and Biblical Institution seems more like a recuperative spa out in the country than a school. The words of student Mary Lapham, penned in 1853, add to the quiet, bucolic impression of restful charm:

> New Hampton is like all country villages with a few houses and stores. The principle [*sic*] street leads past the hotel, seminary buildings, church, and boarding houses. The seminary, white with green blinds, stands beside the church with neat grounds in front. It is a long two-story structure with a cupola where the bell swings, and two piazzas where the girls stand and chat with their arms about each other, or sit with their books and look over their lessons.[45]

While the school's resort-like atmosphere may well have been a factor in Bellamy being sent to New Hampton, larger reasons behind his attendance can be found in its claims to be free "from temptations to *idleness* and dissipation," as well as "saloons, billiard halls, or other temptations to wrongdoing." Though the particular symptoms of what constituted John Haley Bellamy's inclination toward "recklessness" in the wake of his failed courtship were never recorded for posterity, given his being sent to a school that repeatedly touted the absence of such in its advertisements, it seems that we can infer that saloons and billiard halls were large players in the drama. This notion is only strengthened when we consider Bellamy's legendary (and much inflated) struggle with alcohol, a facet of his life to be discussed in greater depth later on in this book.

Relocated to a sleepy rural township nestled in the remote foothills of the White Mountains, far removed from corrupting influences, we can begin to appreciate the logic behind Charles Gerrish Bellamy sending his oldest son to the New Hampton Literary and Biblical Institution. The rationale for his decision is further understood when considering the school's mission statement:

> . . . the objects of this Institution shall be the promotion of Literature, Science and the useful arts, Morality and the doctrines of Evangelical Religion, as revealed, contained and taught in the sacred Scriptures of Divine Authority, being given by the inspiration of God . . . [46]

As its name and mission statement both strongly implied, the New Hampton Literary and Biblical Institution at New Hampton, New Hampshire, was heavily devoted to a religious agenda. From its very founding, the school was ecclesiastical in its outlook. And "all candidates for admission . . . [were] expected to be prepared to give satisfactory evidence of good moral character."[47] As to which religious denomination the school instructed its students, while originally Calvinist Baptist, by 1852 the school had become a Free Will Baptist institution of learning.[48] The fact that the Institute was Free Will Baptist in the character of its organization and instruction is particularly significant, for Bellamy's father was a devout member of that faith.[49]

The spiritual focus of the New Hampton Literary and Biblical Institution was undoubtedly a key (if not *the* key) reason for John Haley Bellamy having been sent off to classes there. By enrolling his son in that school, Charles Gerrish Bellamy sought to distance John from saloons and "temptations to *idleness* and dissipation," while simultaneously imbuing him with the religious zeal and moral fervor necessary to set his life back on course. That this was his prime motive is strongly supported by the fact that the New Hampton Literary and Biblical Institution was renowned far and wide as a place for concerned parents to send young men and women specifically in need of salvation.[50]

A final reason behind Bellamy's attendance may have lain within his father's desire for his son to follow in his footsteps and embark upon a political career. Safford speaks of how "his [John Haley Bellamy's] lack of ambition to develop his 'great political promise' was a disappointment to his father" (MVS: *JHB*, Pg. 4). While this statement can be brought into severe question (and will be discussed in greater depth in the next chapter), a grain of truth may be present, for just as the New Hampton Literary and Biblical Institution was famous for saving the souls of wayward young adults, so was it a launching pad for successful political careers.

The list of students contemporary to Bellamy who attended the institution before embarking upon careers in public service is extensive: Stephen Gordon Nash (class of 1838), justice of the Massachusetts Supreme Court; Benjamin W. Flanders (class of 1836), member of the US Congress and governor of Louisiana; the Honorable H. Merrill Plaisted (class of 1849), member of the US Congress, Maine attorney general, and governor of Maine; George G. Fogg (attended briefly in early 1830s), US senator and US minister to Switzerland; Nathan Clifford (class of 1824), Maine state legislator, speaker of the Maine House, Maine attorney general, attorney general of the United States, and US Supreme Court justice; and "Long John" Wentworth (graduated pre-1836), mayor of Chicago, Illinois, six-term congressman, and "perhaps New Hampton's most colorful and best-known graduate."[51] That such men from New Hampton successfully gained elected office would not have been lost on the politically-minded patriarch of the Bellamy family and surely weighed into his decision to send his eldest son there.

Regarding the nature of his time at the Institution, research conducted by the late Kittery Point historian Joseph W. P. Frost noted that instead of immersing himself within the school's classical, scientific, academic, or commercial courses of study, Bellamy curiously focused upon penmanship, particularly linear drawing, lettering, and decorative scripts.[52] We can surmise that Bellamy's principal instructor at New Hampton was Algernon P. Shattuck, who served as the school's teacher of penmanship.[53] A surviving correspondence sent to his father while Bellamy was a student there includes an itemized listing of his expenses accrued over a ten-week term. From a charge of $22.50 for ten weeks' board[54] (which included room and washing), we know that Bellamy was boarding at the school, since the institution charged "from $2.00 to $2.25 per week" for this service.[55] The cost for his tuition over that same period of time came in at $5.00.[56] Nor was John Haley Bellamy the only Bellamy in attendance at that time, for his half-sister Catherine was likewise enrolled at the institution to become a teacher.[57]

Aside from these scattered facts, precious little is known about Bellamy's time at the institution. Questions of why he pursued penmanship instead of the hard sciences or theological topics are unanswered. And unlike with his apprenticeship under Dockum, we have no idea how well he did or whether the lessons at this school agreed with him.

Emerging into his early twenties after a successful apprenticeship to one of Portsmouth's finest woodcarvers, a vexed, perhaps distraught John Haley Bellamy had presented his family and friends with a worrisome example of a life gone astray. Smarting from an acute disappointment in matters of the heart, Bellamy had seemed to be on a collision course with disaster. By removing him from his familiar environs, Charles Gerrish Bellamy endeavored to shake his eldest son back into life, first by immersing him in meaningful labor and then, when the world financial market declared otherwise, by exposing him to a religious community known for its revivals and soul-saving evangelicalism. He may have also hoped that his woodcarver son would take to a career in politics. John Haley Bellamy's father had responded with alacrity and concern to his son's distressing situation, but he was left to wonder if his ministrations had indeed been successful or if all had been for naught. As the 1850s drew to a close and the American Civil War loomed on the nation's horizon, Charles Gerrish Bellamy was soon going to find out.

The Lancaster Eagle Part 3

IN CHOOSING TO ORNAMENT THE BOW OF THE USS *LANCASTER* WITH AN EAGLE FIGUREHEAD, NAVAL CONSTRUCTOR T. D. WILSON AND JOHN HALEY BELLAMY had elected to do more than make a simple statement of patriotism and national prestige. Bellamy's carving of America's national mascot was nothing less than a direct participation in a tradition nearly as old as civilization itself.

Ever since man first lashed together a few wooden planks, fashioned a sail, and set off across uncharted waters, there has been a constant need to personify vessels as somehow being alive. We give ships names. We invariably refer to them as being female. And until roughly a hundred years ago, it was a common practice to decorate them in a way that sought to exemplify that spark of life perceived to dwell within their timbered frames. The ancient Egyptians were perhaps the first to anthropomorphize watercraft, painting the "Eye of Horus" upon the bows of their reed boats. The ancient Greeks soon followed suit, painting the similar "Oculus" on the front of their triremes. The bows of Phoenician galleys probably sported the first true figureheads, meant to resemble horse's heads. It was a practice that would stretch far beyond the waters of the Mediterranean to the mist-enshrouded fiords of Scandinavia, where Viking vessels became famous for their menacing dragon prows, and onto the high seas as the British lion rode high upon the bow of eighteenth-century English warships when Britannia ruled the waves.

Fig. 3.1 Eagle figurehead upon the bow of the clipper ship *Southern Cross*, detail from Fitz Henry Lane, *Clipper Ship "Southern Cross" Leaving Boston Harbor*: oil on canvas, 1851. [25¼ x 38 in.] *PEM.*

America was not exempt from this practice. In 1689 an Edward Budd and Richard Knight of Boston, Massachusetts, commissioned what was destined to become the first American figurehead, a lion for the bow of the sloop *Speedwell*.[1] That the first American figurehead was a lion—the prime symbol of English sovereignty—should come as no surprise. The lion remained the dominant theme in New England figurehead carving well until the mid-1760s when, on the eve of the American Revolution as an indication of emergent political sensibilities it was dethroned and replaced by less monarchical carved motifs and figures meant to directly represent the vessel's name.[2]

New motifs on the bows of American ships after the Revolution included animals, birds, personifications of Liberty and Columbia, and figures of statesmen, celebrities, historic personalities, and fictional characters aplenty.[3] Independence not only gave rise to new themes and subjects for American figureheads, however. It had also ushered in the emergence of a truly unique American vision in the nautical arts, one where the new country's mythologies and national identity were given voice and brought to life dramatically.

By the time John Haley Bellamy entered his apprenticeship under Samuel Dockum, a revolution in naval architecture dramatically affected the design and aesthetic principles of ship figureheads. Long, slender, and "stripped as bare as Grecian athletes," every inch of the new clipper ships was meant to increase speed as by mid-century a vessel's reputation for swiftness was essential in garnering prime freight rates.[4] Sharp forward angles, hydrodynamic proportions, and the removal of all ponderous, superfluous detail became the general norm for these new vessels and, like the ships themselves, figureheads became streamlined. Designed as structural extensions of the converging lines of the prow, they thrust proudly forward from the bow, facing fearlessly into the salty wind and becoming in the process the most glamorous of American carvings.

Fig. 3.2 S. W. Gleason & Sons, eagle figurehead from the clipper ship *Great Republic*: gilded wood, 1853. [length 61 in.] *MSM*.

Figurehead motifs that suggested speed and flight became popular, an evolution that naturally led to clippers being adorned with eagle figureheads. The eagle proved a fitting and popular choice, with numerous examples of the great raptor created up and down the Eastern seaboard specifically for the purpose of adorning clipper bows. Whether appearing seated in its full glory, [Fig. 3.1] swiftly poised for flight, or merely as a bust, [Fig. 3.2] the national emblem was chosen to represent some of the most famous of all clippers, including naval architect William Webb's *Challenge* and Donald McKay's *Great Republic*.[5]

Thus, by the time John Haley Bellamy began to plan out the *Lancaster* Eagle, figureheads were becoming a disappearing art form. This was especially true for the United States Navy. By the middle of the nineteenth century, the navy had already begun to turn away from commissioning figureheads for its warships, favoring florid yet often generic pieces of carved scrollwork called billetheads in their place.[6] This trend was accelerated during the Civil War, where for economy and expediency American warships, Union and Confederate alike, were commissioned with little, if any, decoration.[7] The movement toward fewer decorative woodcarvings aboard ships continued in the postwar years.

> With the coming into general use of the steel and iron steam vessel in the latter part of the nineteenth century, there was no longer the need of ship carving as before. The old-style decoration of a carved wooden figure did not lend itself to the new craft of steel and iron with the plumb bow. Experiments in decorating vessels of iron and steel with wooden figures only served to strengthen the fact that they were unsuitable and that ship carving in the old manner was doomed.[8]

While the demand for fashioning figureheads steadily dwindled with every passing year, John Haley Bellamy toiled at the twilight of a vanishing artistic tradition seemingly as old as time itself. With tools in hand, he began to carve the very swansong of the shipcarver's craft.

CHAPTER 3

"Missing Years" and the American Civil War

THE IMPRESSION THAT JOHN HALEY BELLAMY left after the dark days of his failed courtship has long had an unfortunate impact upon our perception of him. His behavior during that brief, tumultuous period was undoubtedly troubling to those who knew him; so troubling that for the rest of his life and beyond it continued to define him. Yet in the rich store of Bellamy family correspondences and testimonies, we find that his forced exile to Boston and New Hampton had the desired effect; Bellamy swiftly returned to his old industrious self.

This mischaracterization of Bellamy's personal character may have first come from his half-sister Catherine. As the mother of Moses Victor Safford, she is the most logical source for the unflattering claims in Safford's description of the great woodcarver:

> . . . John failed to meet the family standards in other respects. His lack of ambition to develop his "great political promise" was a disappointment to his father, and although himself of artistic tastes he regarded wood carving as "no sort of a business." John disturbed other members of his family by what they regarded as "sporting proclivities," his fondness for expensive clothes and all the good things of life, but what really got John "in bad" with his family, bred with a stern sense of duty, was John's failure to accept personal responsibility of any sort and his success in ducking any unpleasant task. The family accepted and treated John as an irresponsible artist (MVS: *JHB*, Pg. 4).

This description of Bellamy has proven quite influential over the years as historians unquestioningly echoed such sentiments, "He was a gentle young man, possessing none of the attributes his father deemed necessary for getting on in the world and this caused friction between father and son." While certainly an entertaining and dramatic image filled with pathos and all the makings for great storytelling, it is as unfounded as the claims of his having attended art school. Ironically, it was Moses Victor Safford—the very same man first responsible for describing Bellamy as "an irresponsible artist"—who first called this impression of his uncle into question:

> One result of my attempts to act as John Bellamy's biographer has been to make me feel that the family appraisal of John may need revision, for I note that no man with harmful habits lives to be nearly eighty years old as John Bellamy did that he was never guilty of dishonesty when he had an opportunity to be dishonest, that he was never idle, always doing something, and that he was 68 years old when he received his first annuity payment and had always been self supporting, had dressed well and lived well, and in contrast to his aggressive ambitious father had always managed to keep out of debt (MVS: *JHB*, Pg. 4).

Safford also noted, "Bellamy's artistic genius was never tried by poverty," and that "notwithstanding Bellamy's tendency to indulge himself, he never permitted his fondness for good things to saddle him with debt."[2]

This is confirmed by the many letters exchanged between John Haley Bellamy and his father, the man with whom he had supposedly experienced so much "friction." In all the known letters between father and son, not one single instance has been found where Charles Gerrish Bellamy castigates his son for having chosen to become a woodworker. Nor did Bellamy complain about any paternal disapproval of his trade. Rather, we actually find instances of father and son toiling side by side on various woodworking projects.

Studying the Bellamy family letters, it is obvious that tensions were absent between father and son, at least through the 1850s, 60s, and 70s, the three decades prominently represented in the existing correspondences. Rather than reading accounts that strengthen the assertions that he was "an irresponsible artist," we find John Haley Bellamy offering his father financial assistance "You was very kind in offering to let me have money if I should desire to go to Boston . . . " his father wrote.[3] John also lived at home with his mother specifically so that she could derive income from him. "I had changed my board . . . now shall return home and let mother have the benefit of what I pay out here."[4] Indeed, he and his brother Charles took out a $3,000 insurance policy on their parents' Kittery Point home.[5] Perhaps the most solid evidence of how Bellamy truly got along with his father and his entire family can be found in the following letter. "We are all well" it begins:

> I have just written Charles telling him how we get on &c &c. I now write to tell you that everything continues as when you left. Last night I presented mother with a barrel of flour $11.00 and other things amounting to $20.00 — So you see we are all supplied for four months enough to take us into Spring all OK. So should your job run short come home and we will make a little business in the shop — I shall now lay in some stock and such other things as will enable us to do something in case of a dull time. Stock is all that we ever wanted I could always find the market for my doings and shall now begin to put in proper condition the things to do with as a nest egg in dull seasons. So you can count on a nice shop job at home should things slip up at your yard. I shall begin to work Elisha into the business in such a way as to be of service to him — So count on this for I don't often undertake an enterprise when I have the chance and fail — So set the thing down as a certainty that our shop will be a snug little factory where a day's work can be put in at a profit . . .
> Your Son,
>
> J. H. Bellamy[6]

This is a man far removed from the vision of an artist "in bad" with his family, dodging personal responsibility and having a "lack of ambition. From purchasing groceries for his family and employing his brother Elisha, to offering work to his father, Bellamy shows himself to be the very antithesis of those disparaging descriptions. Claims that Bellamy was a disappointment to his father, and that the two frequently clashed, prove the most unfounded. Indeed, Bellamy repeatedly made employment offers to his father in the autumn of 1873 and the following winter, reminding him that "I shall be able to have work for all of us in case business outside should fail us,"[7] and stating, "if you get through at your yard come home as you will have a good chance to rest and enjoy a comfortable winter."[8] In his father's response to these invitations we discover a man who spoke to his son as a capable equal, as one whom he considered to be the family leader in his absence:

> Dear Son,
> . . . I was much pleased to hear that you had again supplyed [*sic*] the Old Ship with provisions for the voyage, and that you had resolved to furnish stock to be manufactured into some profitable enterprise. Now it is a fact, that without stock we lose much time that might be spent profitable. It is for our interest to keep *all the stock and tools at* home passable, and make that a comfortable and convenient place for business when we cannot make it more profatable [*sic*] elsewhere it is one of the pleasantest places for business in New England

and the cheapest place to live it is true there is no business here at present and we must adopt your method, of making business where there is none, and not depending upon others to employ us, by your method we can be independent and not lyable [*sic*] to there [*sic*] infernal discharges when others think proper.

In union there is strength and if we can keep united and act together as one man, we can soon make that place shine like a new Dollar. I think that Elisha will work to greate [*sic*] advantage there when he gets fairly started. You can do much to put him in the right way. I think he does well for the experience he has had in shop work.

. . . should I leave here, or get suspended I hope you will have something we can make profatable [*sic*] at home. We have had another fearful discharge, and nearly all are cleared out. . . . I thank you for the invitation to come home and lay back a while. But so long as I can make it profatable [*sic*] here it is best for me to stop so take care of all our interests there and write me how things are going on there.

Yours Truly,

Chas. G. Bellamy[9]

The myth that Bellamy was lazy, impulsive, and generally self-centered is not the only legend to have arisen about him over the years. Without supporting testimony or documentation, some authors have chosen to describe him in effeminate terms: "John Bellamy . . . was a 'sissy'," and "a gentle young man..." filled with "naivety and trust..." and possessing a "searching, sensitive soul."[10] Yet, turning to Bellamy family letters and contemporary testimonies, we discover the great woodcarver to have in fact been a rather earthy character, his voice being more akin to that of a salty mariner than an artist with tender, delicate sensibilities. He often refers to his family home in nautical terms, making statements like "we will anchor in the old ship Mansion,"[11] and "So we . . . shall fit the old Pepperrell House out like a Man of War to take us through a long winter."[12] And in

Fig. 3.3 John Haley Bellamy: ambrotype, ca. 1860. *WAF.*

one particularly memorable letter to his father, in describing the political infighting swirling around the construction of a new meetinghouse in Kittery Point, Bellamy's cutting wit and blunt, even coarse, sensibilities shine through:

> The meetinghouse still hangs by the gills . . . Abraham Shaw has got the money all subscribed for the rooster which they are to have placed on the steeple—it is to be of metal and will cost $40—they asked me what it was emblematic of and I told them it was a fit emblem—Emblematic of fighting and fucking. They are now having the fight and how soon they will have the rest I don't know.[13]

From out of the many past attempts to explain Bellamy's character and define his temperament, we find one description of the artist that does ring true, the claim that "in his prime John Haley Bellamy might have served as a model for an old French chevalier" (MVS: *JHB*, Pg. 7). Photographic evidence [Fig. 3.3] shows that Bellamy was still (if not even more) dashing and handsome after his ordeal than he was before.

What he was up to in the years immediately following his time at New Hampton has been a mystery for students of Bellamy. For decades it been said that "from the autumn of 1857 when John was at the New Hampton Institute until after the Civil War when he was working in Boston, we have a gap of approximately eight years. There are no letters or family diaries to fill in this dark area."[14] Fortunately, never-before published documents shed a healthy dose of light onto the missing years of Kittery's favorite son.

While it is probable that Bellamy might have been working as a woodcarver in Portsmouth in January 1859,[15] the first definitive clue as to what Bellamy was up to during this period of his life dates from April of that year. From a fragmentary letter at the Dyer Library of Saco, Maine, we now know that Bellamy was still in attendance at the New Hampton Literary and Biblical Institution early in that year:

> I have several times attempted to write, but there being nothing of interest transpiring, I thought it best to wait until I could get time and opportunity to narrate something pleasing. But after the exercise of much patience, and long delay, I am determined now to begin and conclude a letter

To 10 Weeks board	225	– $22.50
½ Cord wood	166	– 1.66
10 Weeks schooling	500	– 5.00
		$29.16[16]

The spring term at the New Hampton Literary and Biblical Institution commenced on February 8 and concluded on April 21, 1859. Thus, John Haley Bellamy was attending the school during this period and appears to recount his debts at the term's conclusion.[17]

Immediately after leaving New Hampton, we find that Bellamy returned to Portsmouth. If Charles Gerrish Bellamy had wondered if his unorthodox remedy to his son's distressing behavior had worked, he soon found he had no reason to worry. By June 1859, Bellamy launched himself back into the world of woodcarving with gusto, establishing himself in a studio at 17 Daniel Street in downtown Portsmouth, in the same building where his old master Samuel Dockum was operating his business[18] [Fig. 3.4].

> CARVING. —It is gratifying to find any sort of respectable business introduced among us, —especially if that business is a display of skill and good taste. We have been gratified in the examination of some specimens of wood carving executed by Mr. JOHN H. BELLAMY, ship, furniture, sign and frame carver, at his rooms No. 17 Daniel Street. The heads of the Indian and of the ox are well executed, —but the bust of a female is a superior specimen of good workmanship, which would do credit to any artist. Look into his room.[19]

In an effort to advertise his new business venture Bellamy also had promotional cards printed. For almost eighty years they were mistakenly believed to have been printed in the 1880s.[20]) These cards list the many diverse woodworking services he offered, [Fig. 3.5] testimony to the high level of accomplishment and varied experience the young carver had already accrued by this time in his life. Of particular note is that Bellamy identified himself upon the card first and foremost as a "FIGURE . . . CARVER." For years it has been debated whether Bellamy ever made any figureheads beyond his famed eagle for the USS *Lancaster*. While it certainly stands to reason that if he was entrusted by the United States Navy to make a figurehead for one of their warships, he must have had previous experience, some past historians have nevertheless declared that he had not, issuing such decisive statements as, "There is no evidence that he [Bellamy] worked on merchantile [*sic*] vessels or attempted

Fig. 3.4 Gleason's Pictorial, *Market Square, Portsmouth, New Hampshire:* lithograph, July 23, 1853. *PA*. Bellamy's studio was located just around the corner from the very center of nineteenth-century commercial Portsmouth. Daniel Street lies between the two buildings on the right. *DP*

the grand figurative carving that they sometimes required. His marine art was confined to the minor decorative carving that naval quarterboards and gangway boards required."[21] Bellamy's nephew Moses Victor Safford was less definite. In 1935 he felt confident enough to write:

> For a long period of years, beginning prior to the Civil War and continuing until after the opening of the present century, he carved figureheads, ornamental work for sterns, panels for gangways, and sundry other decorations for other parts of naval and mercantile craft, large and small.[22]

Just three years later in 1938, however, a far less certain Safford offered a more cautious statement: "I never knew of his [Bellamy's] carving an old time man or woman ship figurehead although he may have done so . . . So far as I know Bellamy never carved a human face or figure although he may have done so" (MVS: *Eagle*, Pg. 9).

We now know that Safford's first declaration was correct. John Haley Bellamy did indeed craft numerous traditional ship figureheads, as revealed in several new, previously unpublished evidence. The first is from the June 18, 1859, *Portsmouth Times* regarding the opening of Bellamy's studio on Daniel Street; it specifically mentioned that Bellamy had executed three

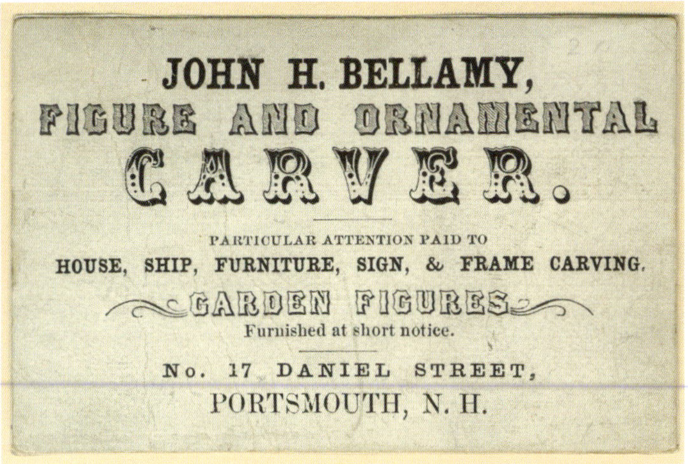

Fig. 3.5 John Haley Bellamy business card, 1859. *WAF*.

figureheads: an Indian, an ox, and "the bust of a female." Confirming that Bellamy had made several figureheads for Portsmouth's merchant vessels his obituary in the *Portsmouth Herald* April 6, 1914, notes, "he carved the figure heads for many of the clipper ships built here when ship building flourished on the Piscataqua river."[23] The *Portsmouth Times* stated bluntly that Bellamy "was also engaged in figure heads when these were in vogue."[24] Another Bellamy obituary shows he strayed beyond crafting figureheads for merchant vessels, having "cut the figure heads for all of the War ships constructed "at the Portsmouth Navy Yard during the American Civil War.[25] Lastly, we have the words of Bellamy himself, who in a letter to his father dated November 10, 1873—a full three years before the navy would commission him to do the *Lancaster* figurehead—casually mentions, "I have just completed and placed on the new ship a head which they [the US Navy] seem highly pleased with—"[26]

It seems highly plausible that the woodcarver from Kittery Point had started carving figureheads while apprenticing at Samuel Dockum's shop and continued to do so during his brief stint at Laban Beecher's old shop in 1857 and from 1859 on through the 1870s. And if the words of Kittery Point historian Joseph Frost are to be believed, Bellamy also made it a point of business to make repairs to damaged figureheads originally created by others.[27]

As to the quality of his creations—while we can easily imagine them as superlative works of the first order, we have at present no way to confirm our suspicions, for aside from the *Lancaster* Eagle no other Bellamy figureheads are presently known to exist. American figureheads were a perishable art form, fashioned from soft pine and subjected to years of storm-tossed seas, snarling winds, and broiling sun. It is most likely that the majority of Bellamy's figureheads—including his Indian, ox, and bust of a female—were destroyed by the ravages of time and neglect. Yet it is also likely that a Bellamy figurehead or two could remain intact, hanging anonymously in a maritime museum or the home of a private collector, silently waiting for the day when it is identified as the handiwork of John Haley Bellamy.

Only one precious piece of his handiwork can be definitively attributed to his time at his 17 Daniel Street studio. Carved in January of 1859,[28] the pine sculpture in question is a delightful likeness of a lion [Fig. 3.6]. The piece is thoroughly Bellamy in its many contradictions: At first glance, the lion appears to meet all the criteria of a crude backwoods folk artist's handiwork, given the overall asymmetrical nature of the piece. The eyes are dramatically different in size, and the center of the beast's upper lip area fails to properly align with its nose. The left portions of the mane differ from the right, an error that is repeated in the brow. And yet for all its failings, the lion stands as a convincing piece of sculpture. At once both threateningly fierce and infinitely charming, its visage exudes an immediate and complicated personality. The deep-set eyes focus with a foreboding intensity upon the viewer, while the soft swish of his mane and almost comical mouth make for a friendly, if not outright cuddly, animal. Amazingly, this layering of traits was accomplished with only a few strokes of the chisel. Its visible gouges and cut marks may lend the lion an air of the unrefined, yet the very fact that such charisma and persona could be crafted with so few strokes

Fig. 3.6 John Haley Bellamy, *Carved Lion "Cat Head"*: pine, 1859. [13 x 14 in.] *OSV*. #20.16.4

speaks of a man who was rapidly becoming versed in some of the finer points of sculpture.

While not the finest carving he produced, Bellamy's lion nevertheless offers a window into his early years as a professional woodworker. Though an unsteady creation possessing obvious flaws, it gives us an appreciation of the rich talent and vision he had already developed by that early point in his formal career. What's more, it reveals that he was producing works specifically for the shipbuilders of the Portsmouth area, for this carved lion's head was not a product of whimsy but rather a nautical device called a "cat head." Used to decorate the end of the heavy timber booms used on the bows of sailing vessels as derricks for raising an anchor, catheads were found aboard merchant and naval vessels throughout nineteenth-century America.[29]

As Portsmouth "had more than a small share in constructing clippers"[30] and Bellamy was making everything from figureheads to catheads with which to adorn them, we would imagine business was fairly brisk for the young woodcarver. Yet it must be remembered that while he began his apprenticeship under Samuel Dockum at a time when "the opportunities for American ship carvers were excellent," and "nearly every coast town along the Atlantic

. . . boasted a thriving shipyard, and whether large or small, with few exceptions, supported a wood carver who did most of the carving for the community . . ."[31] in the aftermath of the 1857 economic depression, shipbuilding and shipcarving activities decreased significantly. Those were lean times for most every shipcarver, and carvers became increasingly inventive in ways to secure a living. "When work on ship ornaments was not available the ship carver turned his hand to mantels, doorways, and other architectural carving."[32] On Bellamy's 1859 business card it shows the young woodcarver was doing exactly this. By listing his ability to work on "HOUSE . . . FURNITURE, SIGN, & FRAME CARVING," Bellamy, like so many other woodcarvers, appears to have been attempting to court whatever business he could.

At that moment, "in the depressed economic climate the artisanal population became more transient . . ."[33] John Haley Bellamy was no exception. In response to the economic pressures woodcarvers in Portsmouth were experiencing, Bellamy closed up his shop at 17 Daniel Street and left the Portsmouth/Kittery area by October 1859 to try his luck along the Boston waterfront. We can easily imagine that Bellamy relocated to Boston on the hunch that he would find more plentiful work there, especially because, as a ship carver, he would have been aware that many a figurehead that had adorned Portsmouth-made clippers had actually been cut in Boston.[34] Moving to the Massachusetts state capital, the second-busiest port along the Eastern seaboard at that time (second only to New York), engaged in shipbuilding, global trade, and manufacturing on a level undreamt of in Portsmouth was a logical choice for a woodcarver trying to find employment. It was also a city where he had once labored (albeit briefly) while at Laban Beecher's former shop, thus assuring Bellamy that he was not taking a leap into the total unknown.

From a brief letter to Frederick Lewis, a former co-worker from 17 Daniel Street days [Fig. 3.7], it is now known he located himself within the shipbuilding neighborhood of East Boston, where famed clipper ship builders Samuel Hall and Donald McKay had situated their enterprises. In the letter, there is a tantalizing glimpse of how the young woodcarver fared in his East Boston life.

> Friend Lewis
> Inclosed [*sic*] you will find $4.00. please do me the favor to deliver the Box to Miss Hacker and the plan to Mr. Hadley, and accept many thanks from your old friend for past-favors until better rewarded. I will send you the $3.00 as soon as convenient. I am well, and doing the same, hope to see you soon. Remember me to all of my old friends, write and tell me how you prosper. Etc.
> I am yours,
> Very Respectfully,
> J. H. Bellamy[35]

Bellamy's reassuring words "I am well, and doing the same" informs us that his days in East Boston generally agreed with him, though it would seem that the money was a touch tight. ("I will send you the $3.00 as soon as convenient.") Regarding whether or not he actually found employment in Boston, Bellamy's obituary from the *Portsmouth Herald* on April 7, 1914, states: "John Haley Bellamy . . . had more than a local reputation, many specimens of his art having decorated large ships built at Boston and Medford, Mass."[36] From this brief testimony, we now know not only that Bellamy found gainful employment in Boston but also worked a few miles upriver in the shipbuilding mecca of Medford, Massachusetts. (In the 1850s the town employed one out of every four shipbuilders in the entire state of Massachusetts.[37]) With a clientele that may have included Donald McKay, Samuel Hall, and Medford's James O. Curtis, it may be that his handiwork would have been no different from what he fashioned in Portsmouth, ranging from full-blown figureheads to minor decorative accents.

A telling clue, which may well allude to the nature of his artistic output when in Boston and Medford, is found in a ship model constructed by Bellamy himself [Fig. 3.8]. While the identity of the vessel is absent from the model and thus will probably never be known, it is significant that this model depicts a

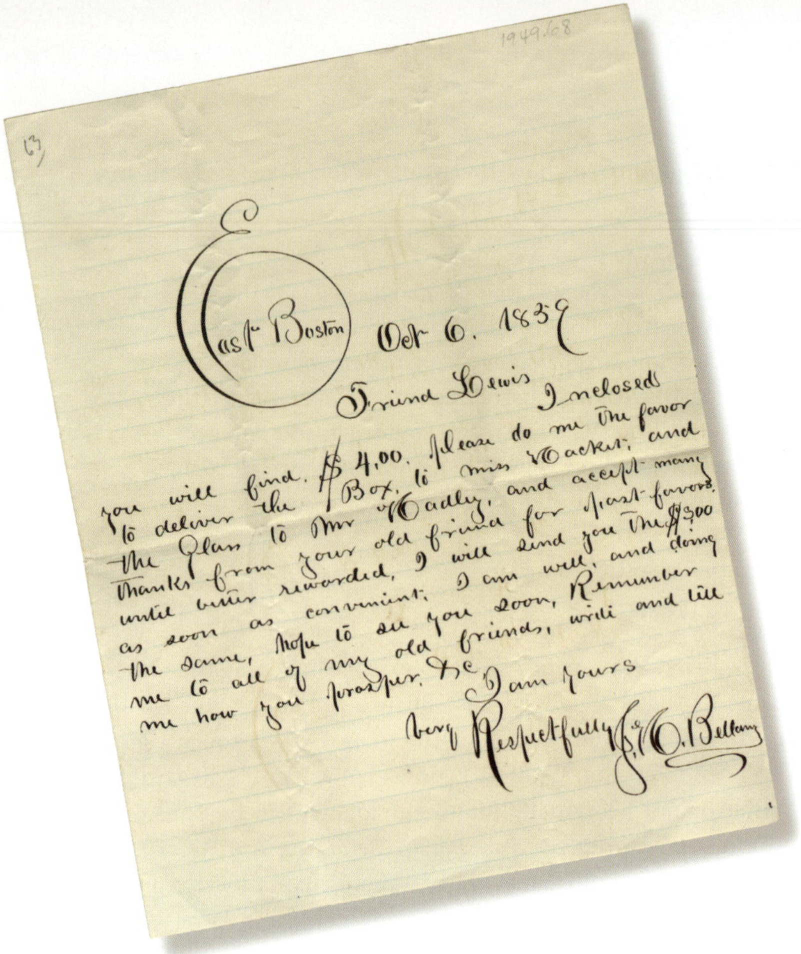

East Boston Oct 6. 1859
Friend Lewis I enclosed
you will find $4.00. Please do me the favor
to deliver the Box to miss Wacker, and
the Plan to Mr Cadly, and accept many
Thanks from your old friend for past favors
until better rewarded, I will send you the $3.00
as soon as convenient. I am well, and doing
the same, hope to see you soon. Remember
me to all of my old friends, write and tell
me how you prosper &c
I am yours
Very Respectfully J. H. Bellamy

Fig. 3.7 Letter from John Haley Bellamy to Frederick Lewis, October 6, 1859. *WAF.*

clipper ship, exactly the type of merchant craft he worked on in East Boston and Medford, as well as in Portsmouth. While the model itself is fairly generic and featureless, the bow stands out grandly with its distinctive eagle figurehead. Completely different from the piece that would one day crown the prow of the USS *Lancaster*, the model is important in that it demonstrates that Bellamy sought to address the challenges of crafting and placing a figurehead upon the extreme bow of a clipper ship while still making the sculpture aesthetically pleasing and structurally sound, imparting a certain air of grandeur to the vessel it represented. By molding the eagle's body so closely to the harsh angle where the bowsprit meets the bow itself, Bellamy successfully emplaced the figurehead in an area notorious for its lack of structural support, securing it to the bottom of the bowsprit instead of having a cumbersome billethead extending out from beneath the eagle for support. By rejecting the billethead, Bellamy also preserved the sense of speed and flight the figurehead was obviously meant to impart. By conforming so closely to the contours of the bow, he minimized the chance of the eagle becoming damaged by wind and waves in this most exposed of locations.

The model also offers us an insight into Bellamy's artistic process. Though at present being the only known Bellamy ship model in existence, there is no reason to assume that it was the only one he ever crafted. It may have also been common for him to craft a model as a means for illustrating to the ship owner exactly how the vessel would appear once the figurehead had been installed. Indeed, it is likely that Bellamy fashioned such basic, rudimentary models whenever he was charged with the task of constructing a ship's figurehead, each model aiding him in envisioning and executing the final product. If these theories are correct, then this ship model is a vital document, offering a precious glimpse of an actual Bellamy figurehead that all indications suggest no longer exists.

Apparently the market for woodworkers in Boston and Medford was little better than in Portsmouth, for John Haley Bellamy left East Boston by June 1, 1860, returning to Kittery Point to live in the Pepperrell Mansion with his family and work as a carver.[38] From his obituary in the *Portsmouth Herald*, we learn that rather than reestablishing his defunct shop at 17 Daniel Street, Bellamy soon "began his work as an expert wood carver at the Portsmouth navy yard during the war of the Rebellion . . ."[39]

Though he returned to Kittery one year before the American Civil War began, it seems plausible (if not likely) that Bellamy

Fig. 3.8 Facing page: (inset) John Haley Bellamy: *Model of unknown clipper ship*: various soft and hard woods, ca. 1859. [23 x 27 in.] with close-up of eagle. *WAF.*

first entered into employment with the United States Navy at that time. Turning to the Portsmouth Navy Yard for gainful employment rather than trying his hand again at running a studio or hiring himself out to private commercial shipyards was a savvy move for Bellamy in several ways. The Portsmouth Navy Yard was unlike so many private commercial shipyards in that it was "a beehive of activity" during the 1850s and into 1860.[40] Moreover, it not only offered better chances of stable employment, it also afforded better work hours. In a commercial yard, work began when the sun rose and ended when it set, meaning grueling sixteen-hour workdays during the height of summer. Employees of a government navy yard were expected to work a standard eight-hour workday in all seasons during peacetime, summer included. Wages were also superior in the government yards. The pay for carpenters, caulkers, and joiners in a commercial yard in Medford, for example, averaged only two dollars a day, a full one to three dollars less than a man could expect to make doing the same job in a government yard at that time, depending upon his level of experience.[41]

For a government posting at the Portsmouth Navy Yard in 1860, connections and political patronage often determined who was hired. Indeed, "politics, winning elections, and the patronage [system] were doubtless important matters" in the hiring process at America's navy yards in the nineteenth century.[42] Charles Gerrish Bellamy's many political connections were likely a prime factor behind his eldest son coming to work for the Navy on the eve of the American Civil War. The Bellamy family patriarch himself was working at the Portsmouth Navy Yard in 1860 as a US inspector of timber, a position he had held since at least 1853.[43] During the war he would continue to labor for the US government, serving as the head of boat building at the Portsmouth and Charlestown, Massachusetts, navy yards (MVS: *Eagle*, Pg. 11). On January 20, 1863 Charles Gerrish Bellamy would also sign on as a subcontractor for work enhancing and improving the defenses at Fort McClary, a coastal fortification meant to defend the Portsmouth Navy Yard, located just down the street from the Pepperrell Mansion (MVS: *JHB*, Pg. 8). For that job, the Bellamy family patriarch saw fit to hire a number of personal friends and family members, including his sons Charles Carroll Bellamy, who worked as the project's foreman for $2.00 per day, and Elisha P. Bellamy, who was the project's inspector and surveyor "for 20 cents per Na. [sic]"[44]

Notably absent from the letters of incorporation drawn up for this project is Charles Gerrish Bellamy's eldest son. The efforts to improve Fort McClary would have profited from the talents of John Haley Bellamy, and yet his name is nowhere to be found there. So, what was Bellamy up to during the Civil War?

The sons of Kittery served their nation with peculiar distinction during the American Civil War. On land, they fought and died in such major engagements as Spotsylvania Courthouse and Gettysburg. Kittery would produce a record twenty-four Congressional Medal of Honor winners during this war, the highest number ever to hail from any one town in any war to date.[45] However, it was at sea that Kittery's greatest contributions to the war effort were felt. "Nearly all of John Bellamy's associates throughout life were or had been in some way connected with the Navy." More than a few of the artist's friends either helped build or actually served aboard the USS *Kearsage*, including during her famous duel with the notorious Confederate commerce raider CSS *Alabama*. Others fought under Admiral Farragut at the naval battles of New Orleans, Port Hudson, and Mobile Bay, while others still "were shot up and cut up at Fort Fisher," and fought "on the monitors at Charleston. One, John Webber went down on the *Cumberland* at Hampton Roads" (MVS: *JHB*, Pg. 11).

At twenty-five, John Haley Bellamy was a prime candidate for service within the Union forces. While his neighbors and friends freely gave themselves to the cause of preserving the Union, Bellamy never enlisted (or attempted to enlist) in the northern army or navy. Nor is there any known record to his having paid to commute his service like his brothers Charles and Elisha did.[46] As an employee of the United States Navy, Bellamy was exempt from military service altogether, given the wartime decision by the provost marshal general that "skilled mechanics and operatives employed in the . . . navy yards of the United States who shall be drafted and on examination, held to service, will not be required to report for duty under such draft so long as they remain in the aforesaid service."[47]

Whether Bellamy was an ardent patriot who sought to serve his country by helping build warships or a draft dodger who escaped service by exploiting a loophole, he spent the Civil War working for the United States Navy. By 1864 John Haley Bellamy is listed in *The Charlestown Directory* as boarding at 29 City Square, Charlestown, Massachusetts, and working specifically as a ship carver down the street at the Charlestown Navy Yard[48]. It is not exactly clear when Bellamy left the Portsmouth yard and moved back to the Boston area or why he chose to do so. Nor is it clear what work he was doing as a ship carver at that time, given that during the Civil War ". . . vessels built for the navies were turned out virtually without decorations of any character. During the war years the craft was practically non-existent."[49]

Whatever the nature of Bellamy's tenure at the Charlestown Navy Yard, there is no doubt that the yard was throbbing with industry. Located in the heart of Boston Harbor, Charlestown Navy Yard numbered among the five "principal establishments" in the US Navy's shipbuilding arsenal before the Civil War.[50] It "made impressive contributions to the Union victory," with its wartime productive output "sometimes equalling [*sic*] if not exceeding, the accomplishments during its entire previous history."[51] As one of 4,995 workers toiling away at the Yard in 1864, Bellamy was part of a massive labor force that successfully constructed seventeen warships, converted twenty-three civilian vessels and forty-three captured enemy watercraft into warships, and carried out the routine maintenance, replenishment, and repairs on existent navy craft.[52]

Yet this era of high-tempo industriousness could not last forever. With the war's conclusion in 1865, the United States Navy immediately embarked upon a program of general demobilization. Consequently, in a total reversal of its wartime mission, the Navy now turned its efforts toward the reduction of its ships and armaments, the disposal of captured and purchased vessels, the discharge and retirement of all volunteers from service, and, in what undoubtedly gave John Haley Bellamy several sleepless nights, the suspension of work on new ships whose keels had already been laid.[53] By May 1865, the Charlestown Yard's staff was cut to the bone and beyond, with its labor force of roughly 3,300 mechanics and laborers in the first five months of 1865 falling to between 900 and 1,100.[54] For those lucky souls who avoided losing their jobs outright, other threats to their job security loomed on the horizon.

Returning Civil War veterans discharged from their service naturally needed to secure employment, and in the postwar job market, where work was scant and pay short, a belief soon arose that men who had fought in the war should be rewarded with navy yard jobs should the veterans desire them.[55] Veterans and veterans' organizations quietly yet insistently maintained that those who had not served should not be hired for navy yard jobs at all, an attitude that bred smoldering hostility among the former warriors now working in navy yards toward men like Bellamy who had been exempted from the draft.[56] With such sentiments running high, it was no surprise when Secretary of the Navy Gideon Welles declared in June 1865, "In the employment of workmen, persons who have served in the navy or army and been honorably discharged will be preferred. Such persons as relinquished their places in the yard to enter the naval or military service during the war will be permitted, if they wished, to resume their position."[57]

With a cessation in shipbuilding, job cuts, an unfriendly work environment, and a movement to replace civilian workers with war veterans, John Haley Bellamy's future looked uncertain, if not outright bleak, given the fact that ship carving was becoming increasingly irrelevant on US naval vessels. Yet within such hardship there was opportunity. Necessity would spur invention within the young ship carver, causing him to become resourceful in the face of adversity. Bellamy would draw deeply from the depths of his talents and find that there was no limit to his creative genius.

THE LANCASTER EAGLE PART 4

BY THE MID-NINETEENTH CENTURY IT WAS FAIRLY COMMONPLACE FOR THE PRINCIPAL CARVINGS ON US NAVAL VESSELS TO FEATURE EAGLES.[1]

Yet it is plain that for John Haley Bellamy, the "Lancaster eagle was more than a ship's figurehead" (MVS: *Eagle*, Pg. 9–10). Rather than merely serving as a decorative motif, "This eagle was conceived as bearing around the world on its broad spread of wings the power and authority of the United States, as further represented by the Lancaster's guns and uniformed crew" (*ibid.*).

Today the idea of employing an eagle to represent American national pride and power may seem obvious, but such was not always the case. When the Continental Congress adopted the American eagle as the emblem to adorn their new nation's Great Seal on June 20, 1782, the eagle was far from being a common American theme.[2] In fact, until the eve of the American Revolution, America was usually represented by images of a half-naked Indian queen, often astride such New World creatures as alligators or armadillos.[3] Given that "she was a woman from Spanish America, a far cry from anything English settlers would have recognized as part of their lives . . ."[4] Congress deemed that this was a poor symbol to represent the United States, and in their hunt for a more fitting emblem decided upon the American eagle as a means to "characterize the new Nation not only to its own citizens but to the world."[5] The US Congress could not have been more right.

> It summarized the determination and the optimism of the naïve new Nation. The Country quite unselfconsciously applied the eagle to its badges, bottles, and butter-molds, as well as its battle flags. It became a popular decorative motif; it perched on cannons, and on shields, was displayed rising, displayed at rest, in flight, in calm security, and as a fierce and furious opponent. It fought with serpents, grasped lightning bolts, aimed arrows of war, and offered the olive branch of peace . . . It even appeared on our stoves and printing presses, and in the copybooks of young ladies at genteel academies during the early nineteenth century.[6]

From the start, the eagle played a prominent role in representing the infant country. When George Washington was inaugurated as the first president of the United States at Federal Hall, on Wall Street in New York, on April 30, 1789, the American eagle presided over the entire affair from atop his lofty perch [Fig. 4.1]. The eagle was also there to greet the new president at every stop during his ensuing four-week tour of the eastern states that year. Yet for all its uniquely American associations, the eagle's formal origins were ultimately European. The European architect Andrea Palladio repeatedly incorporated an eagle motif within his interior décor designs. His influential *Architecture de Palladio*, published in Paris in 1764, was found in the collections of several noted American architects including Thomas Jefferson and Samuel McIntire.[7] Likewise, many of the craftsmen and designers who employed the eagle motif in the early years of the American nation's existence were themselves European:

> Benjamin Henry Latrobe, who dictated the eagles in the new Capitol building, was English, and the carvings produced for him were by Italian sculptors. John Reich, who cut the die for the 1809 half-dollar with an eagle

Fig. 4.1 Amos Doolittle (after a drawing by Peter Lacour), *Federal Hall, The Seat of Congress*: hand-colored engraving, 1790. [16½ x 12 2/3 in.] *Private collection*.

that appeared on our coins until 1908 was German, and the bird's designer is believed to have been Italian. Chief Engravers of the Mint, William Barber, Charles Barber, and George T. Morgan, who were responsible for many versions of the eagle, were born in England . . . [8]

Regardless of its European antecedents, the eagle and the ideals it represented had become a thoroughly American symbol by the time Bellamy began laying out his preliminary sketches and plans for the *Lancaster* figurehead, one fit to soar atop the bow of a US naval warship as it patrolled the world's oceans.

Fig. 4.2 Charlestown Navy Yard, with USS *Wabash* lying "in ordinary" in the center of the picture: photograph, ca. 1870. *HNE*.

CHAPTER 4

The Postwar Years

THE CHARLESTOWN NAVY YARD WORKFORCE was reduced by more than two-thirds immediately following the American Civil War and for those who remained was a less-than-friendly workplace. Job security was almost nonexistent. Fear of future layoffs and the incessant jockeying for jobs and wielding of political influence by those who remained bred clannishness and paranoia. Yet, in what stands as a prime testimony to how highly valued were his talents and to his determination to work for the navy (and the possible scope of his father's political influence), Bellamy remained at the Charlestown Navy Yard, defying budget cuts and popular hiring practices. Why did Bellamy tenaciously clung to his employment at the Charlestown Yard rather than returning to self-employment or work in a civilian yard? In comparison to the work conditions most Americans enjoy today, one may wonder just why Bellamy found working for the navy so desirable. Undoubtedly, humane work hours and a steadier wage—the very factors that attracted him to the navy—played a role in his decision.

Like all his fellow Navy yard workers, Bellamy's days were ruled by the ringing of the yard bell, which marked the beginning and end of the workday and the beginning and end of the noontime lunch break with its dolorous clang. As a military installation, customs, like the muster, inevitably trickled down into the organization of the civilian workforce. Each day before work began, after the noontime meal, and at the conclusion of the workday, yard employees were mustered in military fashion, the names of those in attendance being carefully recorded. Though the workday had spanned sunrise to sunset during the war, it now returned to a standard eight-hour shift, with yard employees working four hours in the morning and four in the afternoon, the two sessions split by an hour-long lunch period to break up the day's monotony. The workweek itself was six days long, Monday to Saturday, interrupted by the occasional unpaid holiday. Pay came only twice a month and constituted the only responsibility the navy held toward its employees. Retirement, health, and life insurance benefits were nonexistent.[1]

Fig. 4.3 John Haley Bellamy, *Initial schematic drawing for* USS *Wabash Sternboard*: pencil and paper, ca. 1865–71. *PEM*.

Fig. 4.4 Sternboard of the USS *Wabash*, serving as a receiving ship at Charlestown Navy Yard: photograph, ca. 1875–1912. *NHHC*.

This new infusion of work put John Haley Bellamy's tools and talents into action in a grand fashion. Unlike during the Civil War, when no specific example or mention of Bellamy's handiwork is presently known, one superior and elaborate piece of woodcarving he executed during this postwar period has come to light. Drawn in his own hand upon the back of a poem penned by Bellamy himself, is a preliminary sketch of a sternboard decoration for the warship USS *Wabash*. A 4,808-ton steam screw frigate that had served admirably in the Blockade of Charleston and assaults on Hatteras Inlet, Port Royal, and Fort Fisher during the Civil War, the *Wabash* had been deactivated and placed "in ordinary" at the Charlestown Yard in February 1865.[3] In October of that year, the *Wabash* was taken out of ordinary to undergo a major overhaul, [Fig. 4.2] at a snail's pace that lasted until 1871.[4] At some point during the extensive rebuilding, Bellamy's woodcarving talents were called upon.

Though the final product was far more elaborate than the sketch, the identity of its creator is hardly in doubt, given that the *Wabash* [Fig. 4.3] was undergoing repairs at the Charlestown Yard at the same time Bellamy was working there. Bellamy was one of the few (if not the only) ship carvers present. The similarities and differences between his initial sketch and the final product are evident in photographs taken of the *Wabash* both

Bellamy's tenacity ultimately paid off. Though only a shadow of its wartime performance, work eventually began to trickle back into the Charlestown Navy Yard:

> In the years 1865-1869, much of the work at the yard consisted essentially of postwar activities. Such work included sale of surplus auxiliary vessels; acceptance, servicing and outfitting ships built by private ship works under wartime contracts; and completion of new construction begun in the yard during the hostilities. In addition, the yard was involved in the construction of several vessels, the keels of which were laid after Appomattox, and there was a fair volume of ship repairs.[2]

immediately after her overhaul and in later years when serving as a floating barracks in Boston Harbor. [Fig. 4.4] From the start Bellamy envisioned the sternboard as sporting a shield in its center, serving as the focal point of the decorative piece. Instead of the simple device he had first sketched, the final product was a far more involved emblem. In the final product, intricate vegetal scroll work akin to other Bellamy carvings surrounds the shield, a detail Bellamy made a passing reference to in his initial drawing with a few amorphous scribbles, proof that he had yet to settle upon a definite design when he first put pencil to paper. Also present is a lone star sitting beneath the sternboard, which does not appear in the sketch; the name "Wabash," proudly emblazoned across the top in the original draft, is inexplicably missing.

It has been said that Bellamy's work at the Charlestown Navy Yard "attracted such favorable attention that a public exhibition of several of his carvings was held in Boston."[5] However, the navy did not regularly employ ship carvers in the years following the American Civil War. Although the names of carvers appear occasionally on navy payrolls and in special reports "most of the carvers who worked for the navy maintained their own shops outside the Navy Yard, where they did work for commercial ships, often carrying on an extensive business."[6] Navy ship carvers were at best part-time employees who worked on a contractual basis while simultaneously carrying on "an extensive business" beyond their government duties.

John Haley Bellamy was no exception to this rule, a factor that goes far in explaining his other activities while employed at the Charlestown Yard. Rather than operating his own woodcarving studio, however, Bellamy began an altogether different venture. According to Moses Victor Safford:

> It would appear that, when in need of funds, instead of attempting to stimulate the market for his carvings, he preferred to turn his attention to quite unrelated fields, to the creation of ideas that might be copyrighted for advertising purposes, and to the improvement of mechanical devices and tools. He was gifted with remarkable mechanical ingenuity. Friends helped him to dispose of his inventions and protected his interests. Most of his contrivances he sold outright; for others he was paid small royalties.[7]

Safford's claim that his famous uncle was talented in matters of invention was cited time and again at Bellamy's death in 1914. That he "was of an inventive genius and in early life designed many patents from which others realized substantial revenues,"[8] "prolific as an inventor, having taken out nearly 30 patents on various articles, some of which are today regarded as almost necessities of life,"[9] and "a man of genius . . . [who] received many patents from the government for inventions,"[10] ran throughout his obituaries, driving home the point that mechanical inventions were a central focus of his life and career.

Bellamy was not alone at the Charlestown Navy Yard in being of a resourceful mind. Given the heavy mechanical principles that were applied every day in the Yard's many diverse operations, more than a few of the men working beside Bellamy tried their hand at invention, creating everything from "a special truss to prevent the chaffing of top masts," to "an improved flier for use in ropemaking."[11] Yet the drive to invent appears to have been a deep, almost primal compulsion for the woodcarver from Kittery Point. He claimed, "I can always think up something better (MVS: *JHB*, Pg. 5)" and it was rightly said that Bellamy had "a whirling brain."[12] When busy in his studio, "Now and then he would stop carving to jot down on paper which he kept for the purpose ideas which might occur to him as being of possible future use either in connection with his carving or writing" (MVS: *Eagle*, Pg. 8–9).

To better capture the sudden ideas and innovations that leapt into his head, Bellamy designed and crafted with his own hands a unique writing slate [Fig. 4.5]. Fashioned from pine and walnut, he used it to record the many ideas that jarred him from his sleep in the middle of the night. Without having to waste time fumbling for a lamp or candle and lighting it, Bellamy only needed to take hold of the pencil attached to the device with a string and find the open space cut into the face of the writing

slate. With the beveled edges of the open space and the groove in the wood immediately below it to guide his hand, Bellamy could pen his thoughts, ideas, and poetry down on paper in pitch darkness with ease. Of course, with such a small writing surface to scribble upon Bellamy needed a constant supply of fresh paper. This problem the artist remedied with a simple turn of one of the knobs on the side; inside the slate was a mechanism of true Yankee ingenuity in its simple, blunt effectiveness [Fig. 4.6]. Mounted on spindles, long sheets of paper could easily be wound along as Bellamy wrote, rolling away under his pencil; a wooden crosspiece behind the paper acted as a writing surface and prevented the paper from being pierced by the pencil. Metal clips guided the paper along smoothly, while spare rolls lay at the ready in separate compartments, should the paper run out.

While it was the drive to invent that caused him to take pencil in hand and scribble down his imaginings, pure financial need served as the prime motive behind Bellamy's forays into patenting and marketing his inventions. From a letter written by Elisha Bellamy during the drastic government cutbacks of April,

Fig. 4.5 John Haley Bellamy, *Night Slate*: pine, paper, and metal, n.d. [8 x 7½ in.] *MM*.

Fig. 4.6 Interior view of John Haley Bellamy's *Night Slate* invention. *MM*.

1866, a time when he was working beside John in the Charlestown Navy Yard, we gain an intimate picture of the low pay and scarcity of work that drove John Bellamy to sell his inventions. Elisha confides to his brother Charles: "Well I'll tell you I have got a room here and am carving that is to say I am trying to carve. I can cut enough to pay my bills and that is doing well for me."[13] He also goes on to explain that the Navy Yard " . . . can get a plenty of carpenters here for 250 [$2.50] per day and they are glad to work for that I tell you times are hard here and there is no prospect of them being any better but we have got through the winter that is one consulation [*sic*]."[14]

John Haley Bellamy's inventions ranged from the simple to the complex, addressing the needs of everyday people and captains of industry alike. Some of the practical devices the woodcarver created were a "self-moving arm,"[15] an improved hauling device,[16] and a new and improved mode of constructing a check-rein, designed specifically "for the prevention of misery inflicted on horses by the use of the common leather strap."[17] Like so many other facets of his life, Bellamy's career as an inventor has been ripe fodder for imaginative retellings and dramatic, ultimately unfounded, claims.

Regarding Bellamy's business savvy, some claimed: "He had a childlike innocence which made him a target for unprincipled associates who took advantage of him. Sometimes his naivety and trust swayed him in directions contrary to his own interests."[18] Yet, this sensationalistic description of Bellamy simply does not stand up to historic fact. From the volumes of contracts in Bellamy's own hand that today can be found in the Peabody Essex Museum's Phillips Library, it is quite obvious that he went to great lengths to protect himself in selling his inventions. He had the foresight to have his business documents notarized and signed by witnesses, and on specific occasions drew up multiple copies. These agreements are anything but inequitable toward him, securing generous payments and fair interest percentages for the artist/inventor. Not only was Bellamy skilled in protecting his business concerns, but it has been authoritatively stated, "Friends helped him to dispose of his inventions and protected his interests."[19] Despite the assertion that he was the possessor of a "childlike innocence," a letter to a Lois M. Faxon shows Bellamy a pit bull in such matters, more than capable in defending his interests:

> To Lois M. Faxon of Boston in the Commonwealth of Massachusetts
>
> I hereby notify you that I do not regard myself bound or held under or by virtue of an agreement entered into between you and me in writing and dated December 14, 1868. You having for a long time failed to keep, observe, and perform your part of said agreement, and thereby having rendered it impossible for me to perform said agreement on my part.
>
> I shall hold you liable for all damages sustained by me for reason of your non performance of said agreement.
>
> And you are further notified not to make any further contracts or agreements in our names under said agreement, which as to all future matters is hereby revoked.
>
> Witness my hand at Boston May 31, 1869
>
> John H. Bellamy[20]

Far from being bluster and empty threats, this letter brought Lois Faxon to terms; she released and discharged Bellamy from "all claims, accounts, and demands" by November of that year.[21]

Of all the inventions that the woodcarver from Kittery Point patented during his time in Charlestown, perhaps none were as lucrative as those he designed for America's Masonic community. As early as 1866, John Haley Bellamy conceived a number of carved wooden frames, clock cases, and what-not shelves replete with Masonic symbols to sell to members of that fraternal organization. Bellamy was far from the first to craft decorative furnishings with such imagery. In America the tradition of garnishing home furnishings and implements with Masonic symbols went back at least to 1769 when artisan H. W. Stiegel fashioned a stove plate inscribed with the Order's sigils.[22] After the American Revolution, Masonry increasingly began to

Fig. 4.7 Titcomb & Bellamy, *Masonic frame*: walnut, ca. 1866–72. [31½ x 22 in.] *SRMM / DB.*

Fig. 4.8 Titcomb & Bellamy, *Masonic frame*: walnut, ca. 1866–72. [31½ x 22 in.] *KHNS / DB.*

hold an influence on the American decorative arts, with chairs, tables, dishes, glassware, woven coverlets, and stoneware jugs and crocks just a few of the everyday items that were embellished with Masonic motifs.[23] This fashion craze enjoyed wild popularity, along with the Order itself, until 1826, when the scandalous alleged kidnapping of William Morgan by the Masons of Batavia, New York, began a wave of anti-Masonic sentiment that led to a sharp decline in Masonic membership.[24]

If Masonry languished in America after the 1830s, the Civil War breathed new life into the storied fraternal organization, for both Union and Confederate men began to enter Masonic military lodges in large numbers. It was a trend that continued into peacetime, as veterans returning home from the conflict popularized the organization among the civilian population.[25] With this widespread resurgence in membership, Bellamy's pieces were virtually guaranteed healthy sales.

Bellamy's inventions were also completely in keeping with Victorian tastes. Rather than furnishing their homes with an eye toward beauty or simple utility, Americans of the nineteenth century were often prone to decorate their dwellings "according to aesthetics that was more concerned with the moral, social, or intellectual ideas implied in the styles of objects."[26] This was especially true in regard to Masonic-themed furnishings like Bellamy's, for in Masonry "every character, figure, or symbol . . . possesses a moral reference and inculcates the practices of moral and social virtue."[27]

Highly decorative and sure to appeal, Bellamy's creations were nevertheless designed for practical uses. The frames, which over time were often converted to hold mirrors, were originally intended to proudly display a master Mason's membership certificate [Fig. 4.7] or a Masonic lodge's charter, while the shelf clocks were offered for use in both the home and lodge.[28]

It has often been supposed by collectors and dealers that Bellamy was himself a Mason given the working knowledge of Masonic symbolism revealed in his carvings. No record exists to confirm these suspicions however; the Grand Lodges of Massachusetts, New Hampshire, and Maine are all in agreement that Bellamy was never initiated into the Order. While Bellamy may have learned the significance and proper usage of Masonic symbolism from Jeremy Cross' *True Masonic Chart* and other popular publications then in circulation,[29] it is also possible (if not probable) that he may have come to understand Masonic symbols through his old teacher Samuel Dockum, a Mason who may have done work for Portsmouth's Pythagoras Lodge No. 33. Regardless of how he became knowledgeable in the Order's symbols, Bellamy's works stand out not only in familiarity with Masonic emblems but for the clever ways he assembled their incongruent, disparate shapes into forms that are aesthetically pleasing.

The latter decades of the nineteenth century were witness to the increasing dominance of mass production and "fewer items with Masonic symbols of this period are really of a quality to be considered as decorative arts."[30] Yet John Haley Bellamy's Masonic frames, clocks, and shelves, by combining art and industry, seem an exception to this trend. It must be noted that John Haley Bellamy did not craft each and every one of the Masonic pieces himself. Writing to his father in March 1866, he mentions that his brother "Elisha is doing well—making the Masonic brackets. He will doubtless be [-?-] with samples in a few days, or a week."[31] Two years later he laments to his father, "I shall not be able to get home tonight as I have two men on models which they do not understand,"[32] while in a letter of September 1872 sent to his brother Charles, Bellamy refers to his father having labored for a time in the Charlestown shop. Bellamy thus informs us that he had established a workshop independent of the Navy Yard and had taken on assistants (both kin and strangers) to help him turn out his Masonic creations. Analysis of the pieces themselves confirms this, for while Bellamy may have conceived and patented them, these articles were not all carved by him.

A comparison between two frames sporting the same decorative design [Fig. 4.8] betrays the hand of two or more different artists at work. Figure 4.7, with its rich attention to detail and sensitive handling, is immediately evident as the handiwork of a master woodcarver. From the famous "all-seeing eye" device at the frame's pinnacle to such minor touches as the lapstraked hull of the ark in the bottom right corner and the purely decorative marks along the bottom bracket, a uniformly precise, steady hand can be found at work throughout the piece, assuring us that this item was crafted by Bellamy and Bellamy alone. The second example, however, is a far more complicated affair. A quick glance informs us that the level of attention in the first frame was not lavished on this piece. The lush, luxuriant sense of texture found in the first example (particularly along the top and in the eye) is somewhat lacking here. The carving has less life to it, appearing a touch flat. Marks such as those found upon the carpenter's square and the bottom bracket are not as deep and precise and appear to have been more of an afterthought. Discrepancies in the design also arise when comparing the trowel at bottom center and the globe atop the right-hand pillar. The

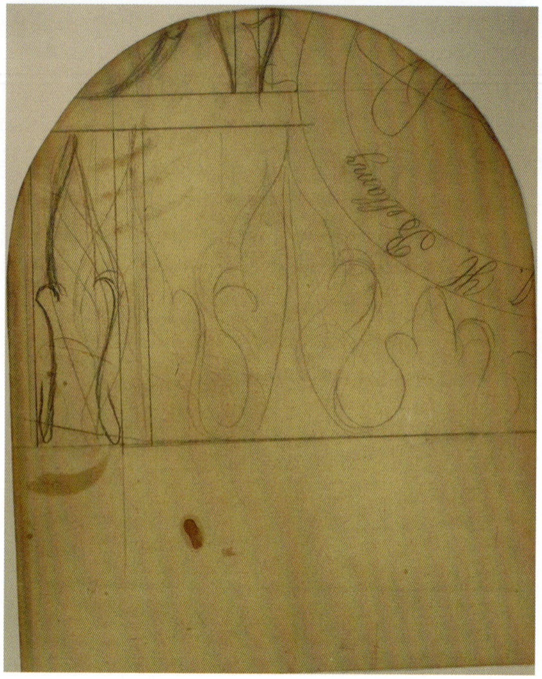

Fig. 4.9 Backing board for a Titcomb & Bellamy frame, with John Haley Bellamy's signature and design notes: pencil on paper, ca. 1868. [17¼ x 14 in.] *SRMM /DB.*

first frame of the trowel, ingeniously fashioned into the form of a shield, bears the initials J. G. C., the cardinal points on a compass, and a detailed handle, while the second frame is plain and unadorned save for a few cuts along the edge to offer a sense of definition to the shape. The globe atop the right-hand pillar in the first frame features a banner with various sigils wrapping around it that stretches diagonally toward the upper right shoulder of the globe, while the same globe in the second frame stretches in the opposite direction.

Why are these differences in detail and handling present? Two complementary reasons exist. The first is time to carve. Though no definite number of employees is known, all available evidence suggests Bellamy had but a handful of assistants in his shop, perhaps two to four or more, depending upon just how busy they were at a given time. From the volume of Bellamy Masonic frames known to exist and the number of states in which they were sold, they enjoyed healthy sales in several parts of the country. High demand and small staff requires pieces to be turned out quickly in order to meet production deadlines, with the finer decorative nuances being an obvious casualty. In such a scenario we can see the first frame as having been produced at a time of slack demand, when enough time existed to go the extra mile and devote more attention to the piece, while the second frame was fashioned during a period of high demand, causing the above-mentioned discrepancies.

The second reason behind these differences is that different artists were at work on the pieces. While the first frame bears the unmistakable handiwork of Bellamy, the second bears the hand of at least two different carvers. The crest along the top appears to have been executed (albeit hastily) by Bellamy himself. The same may be claimed for the two globes and pillars, but in regard to the bottom bracket, a close examination reveals this piece to be too shallow and uninspired in its carving to be Bellamy's handiwork, even under the most rushed of conditions. Unlike the crest along the frame's top, where the edges have been rounded and softened, here the edges are sharp and cutting, having none of the finesse Bellamy instinctually placed into his work. Consequently the symbols take on a static, two-dimensional "cookie-cutter" quality.

If these explanations are correct, Bellamy was indeed using small factory assembly methods within his workshop situated at 11 City Square, Charlestown. Rather than handcrafting each piece from start to finish himself, Bellamy assembled his frames and clock cases in a fashion quite common to furniture makers of the late eighteenth and nineteenth centuries (his mentor Samuel Dockum included), parceling out various tasks to his assistants while participating in the process only at certain key junctures. Through close examination of the frames, we can now see their construction began with one man taking broad, thin pieces of walnut and cutting out the needed designs with a mechanical scroll saw, a paper pattern of Bellamy's design serving to guide the saw and assure consistent quality. This part of the operation would have been fairly monotonous, with an assistant repeatedly cutting the same design over and over until the requisite number for an order was reached. Once cut, the faces of these pieces would be carved by hand, though scroll saws were employed for certain pieces where mechanical precision was essential, such as the pediments at the bottom of the two columns and the molding. And even where the wood was incised by hand, paper patterns would have been used to assure general uniformity among the duplicate pieces.

In this step in the process Bellamy himself played a large part, personally carving out many of the faces to his frames, shelves, and clocks. Proof of this can be found not only in the presence of his superior signature handiwork on the face of a frame, but in at least one known instance Bellamy's signature is found upon the back, scrawled amid a flurry of design notes for other possible furnishings [Fig. 4.9]. Yet, we now know, he used assistants here as well. As this work was both delicate and crucial to turning out a successful product, it can be argued that Bellamy would only have entrusted this task to his brother Elisha as he would have understood Bellamy's methods and approaches to woodcarving better than a worker off the street. After the pieces were carved, the pieces were sanded, shellacked, assembled, and finally packed for shipment, all functions that in times of high demand may have necessitated taking on more laborers. Joining the various parts together on a frame was done with the aid of splines, a tongue-and-groove method that, while time-consuming, assured greater durability and longevity than would have been achieved with nails.

Bellamy patented at least six different Masonic frame designs while in Charlestown.[33] While true to their central theme, the designs themselves stray across a wide range of aesthetic approaches, ranging from stiff, restrained pieces that are quite similar in their symbols and arrangement of the previously discussed examples [Fig. 4.10] to loose, organic compositions that feature a different array of Masonic symbols [Fig. 4.11]. In one curious instance Bellamy went so far as to include an emblem that, while holding no immediate Masonic association, reflected his own patriotic sensibilities and certainly appealed to those of his customers, the American eagle [Fig. 4.12]. Yet regardless of the particular symbols and compositions employed in creating his frames, all of Bellamy's Masonic designs are unified in their employment of Gothic aesthetic principles.

At the heart of Masonic philosophy is the notion that God, the Creator of the Universe and all that is found within it, is "the divine architect." With this underlying tenet as their guide, Masons believed that by understanding architectural precepts one could ultimately come to gain an understanding of the Divine and how it functions. Such attitudes naturally fostered an interest in architecture and decoration within the Order, an interest which by the time Bellamy was working in Charlestown had easily merged with Victorian tastes for historical revivals of Gothic, Renaissance, and Italian architecture.[34] Whether a wild tangle of grapes, roses, and leaves or a clever arrangement of Masonic symbols that together simulate armorial devices, the delicate curves, soft lines, and medieval qualities of Bellamy's compositions place them firmly within the domain of the Gothic Revival. It is much the same regarding Bellamy's Masonic clocks. Indeed, whether sporting a Romanesque arch along the top [Fig. 4.13] or having the Masonic square and compass form a stylized spire, [Figs. 4.14 and 4.15] Bellamy's clocks appear more like small chapels from the Middle Ages than timepieces hailing from the Industrial Revolution.

If Bellamy's clocks are reminiscent of medieval churches, then his Masonic what-not shelves are certainly akin to miniature votive altars. Small wooden shelves meant for the display of curios and bric-a-brac in the home, these are perhaps the rarest and most unusual of Bellamy's Masonic offerings. Given their overall rarity, it has been postulated from time to time that Bellamy custom-made these articles, each piece being a strictly one-of-a-kind affair. Yet examination of one particular example in the collection of the Kittery Historical & Naval Museum [Fig. 4.16] strongly alludes to their having been assembled in an industrialized fashion identical to his frames.

Fig. 4.10 Titcomb & Bellamy, *Masonic frame*: walnut, ca. 1870. [31 x 19 in.] *SRMM / DB*. This piece is an unusual example of Bellamy's Masonic frames in that it was carved in relief rather than cut completely through to produce a filigree.

Fig. 4.11 Titcomb & Bellamy, *Masonic frame*: walnut, ca. 1870. [27 x 18½ in.] *SRMM /DB*. Replete with exquisitely carved grapes, roses, leaves, and stylized flowers, this frame stands apart from Bellamy's more symmetrical Masonic offerings. Despite being an atypical "organic" composition, this frame does sport Masonic symbols and remains one of the true masterpieces from this point in Bellamy's career.

Fig. 4.12 Titcomb & Bellamy, *Masonic frame with eagle*: walnut, ca. 1866–72. [27½ x 22 in.] *SRMM / DB.*

Fig. 4.12a Titcomb & Bellamy, Masonic Knights Templar frame: walnut, ca. 1866-72. [33 x 26 x 2 in.] *Grand Lodge of Texas*.

Fig. 4.13 Titcomb & Bellamy, E. N. Welch Clock Co., *Masonic Royal Arch shelf clock*: walnut, glass, brass, steel, ca. 1870. [15¼ x 16 x 4¼ in.] *SRMM /DB.*

Fig. 4.14 Titcomb & Bellamy, E. N. Welch Clock Co., *Masonic shelf clock*: walnut, glass, brass, steel, ca. 1870. [19½ x 13¾ x 3¼ in.] *KHNS /DB.*

Fig. 4.15 Titcomb & Bellamy, *Masonic shelf clock with Knights Templar symbols*: walnut, brass, pine, ca. 1867. [16½ x 15 x 3 in.] *SRMM /DB.*

Fig. 4.16 Titcomb & Bellamy, *Masonic what-not shelf with Knights Templar symbols*: walnut, ca. 1866–72. [19 x 11¾ x 5¾ in.] *KHNS / DB.*

Fig. 4.17 Close-up of Masonic *G* upon Titcomb & Bellamy's *Masonic what-not shelf with Knights Templar symbols*: walnut, ca. 1866–72. *KHNS / DB.*

Fig. 4.18 Close-up of Maltese Cross upon Titcomb & Bellamy's *Masonic what-not shelf with Knights Templar symbols*: walnut, ca. 1866–72. *KHNS / DB.*

Like the frames and faces of his clocks, the ornate back of the shelf was originally cut from a single piece of wood with a scroll saw, as were the shelf itself and the bracket beneath that supports it: a sure sign of mass production. And as with the frames and clock faces, the front of the shelf and the supporting bracket were hand carved, apparently by both John and Elisha Bellamy, given the presence of dissimilarities such as the expertly carved *G* set in the top center of the piece [Fig. 4.17] and the sophomoric Maltese cross situated in the bracket supporting the shelf underneath [Fig. 4.18]. Another allusion to mass production is the pillars on the left and right. Sporting different capitols, pedestal bases, and fluting, the two pieces are a complete mismatch, an anomaly that could only be the result of a mistake made in the rush to fill an order on time, for no other known Masonic shelf (or frame or clock) by Bellamy bears non-matching pillars.

Given the complex, intricate, texturally rich façades Bellamy's Masonic-themed clock cases, frames, and shelves sport, it seems only natural for them to have been painted in a wide array of vibrant colors, if not partially gilded as well. Indeed, they nearly *scream* to be colored, yet as the reader can plainly see from the examples featured above, Bellamy's Masonic carvings were overwhelmingly varnished or left plain. (That they may have once been painted and/or gilded and stripped at a later date is refuted upon close visual inspection; no trace elements of paint or gilding can be detected upon them.) Why Bellamy did not paint his Masonic creations most likely involves the very thing that also demands that they be so decorated, the richly ornamental qualities of the work. With their many trappings and trimmings, the application of paints to these carvings would have been a taxing, time-consuming process, especially when considering the need for careful, steady application amid the tangled surface of the carvings and the need to be painted in stages to permit for drying time. Consequently, Bellamy would not have been able to meet his production demands had he painted them, so he merely varnished them before they left his shop.

Fig. 4.19 Titcomb & Bellamy, *Painted Masonic what-not shelf*: painted walnut, ca. 1866–72. [22 x 12 x 6½ in.] *Private collection, AKA / GA.*

There is, however, one Bellamy what-not shelf that offers us a startling glimpse of what his Masonic pieces would have looked like had he had enough time to so adorn them. Deftly cloaked in rich hues of red, blue, white, gold, and green, the piece in question is to date the only known example of a Bellamy Masonic piece to be fully painted [Fig. 4.19]. By visual analysis, the paint is from the same period as that of the shelf's construction, a revelation that, coupled with the fact that the paints were applied with a consideration toward balance and tonal value that conforms to known color schemes used within the Bellamy workshop in later years, verifies that the paint treatment was most likely a product of the master woodcarver's and not applied by other hands at a later date.

If this shelf was indeed painted in Bellamy's Charlestown shop as the evidence strongly suggests, we are left to ponder why he went to the trouble of having this example painted when so many others were shipped out the door unembellished. The possibilities are legion: this shelf may have been painted on special order or as a gift to a particular individual or Masonic temple. That an assistant painted it for personal use is also quite possible, especially if he was a Mason himself. Alternatively, Bellamy may have had this (and other) examples of his Masonic handiwork painted specifically for display purposes in the window of his shop to garner the interest and attention of passersby. Or perhaps it is one of a handful of samples his business agent used when peddling his Masonic carvings in other states. In any event, this shelf stands as a striking demonstration of how Bellamy envisioned his Masonic-themed pieces should look as a finished products and may very well stand as the rarest and most unique documented piece from this time in his career.

The events that led to this artistically productive, financially lucrative era in John Haley Bellamy's life began in the early months of 1866, as evidenced in an unpublished letter from the artist in Charlestown written to his father on March 3:

> We are all well. I received your letter and should have answered it before but have been very busy. Elisha is doing well— making the Masonic brackets. He will

doubtless be [-?-] with samples in a few days, or a week. Since I produced the signs for the brackets I have made a frame that is meeting with great favor, so much so that I have made application for a patent and shall probably receive it soon.

There is a firm at the City (Boston) that is now engaged in making the frames. They, the firm, have changed their place of business, taken a large shop and are desirous that I take charge of this part of their business. They now pay me (or will as soon as they sign the contract), one third of all that is made and they have offered me one third, or an equal partnership in the business if I will leave the Navy Yard, take charge of the shop and allow all my designs to be used by them. I am yet undecided as to what I shall do, but think as matters look now, that I shall leave the Navy Yard before long.

There have been two parties who have offered me money to start the business myself, desiring to be a partner, but I did not like the men, thinking that it would be better for me to go in with some one in the business who had withstood the storms.

You will hear again from me soon then I shall know what action I shall take. As matters look now the future is bright. There is one thing I can say as to this work of mine. It is original with me and never was known or heard of until I produced it.

One man employed at the Navy Yard on seeing it and he being "Free Mason" desired that I should allow him to be an agent if we had any. He was so anxious that I took him to the City and introduced him to the makers. They thought favorably of the proposition so I cut him up a sample. He took it out for one day and was so pleased with his success in placing orders that on the following day he left the Navy Yard and now has an office at the City and is acting as our agent for the State of Mass. He was receiving when he left the Navy Yard the same pay as myself.

There is another agent who will start for the West in a few dats. [*sic*] Our firm on Cornhill are so confident of success that they went to obtain the State of New York for their agent. How we will fix the matter I cannot say at present. If I get the patent, as seems certain now, I may sell out some of the States to other parties.

I shall write soon— so good-by for the present. Give my love to all and write me soon.
Your loving son,
John[35]

Bellamy's own words indicate that his Masonic designs were an instant hit and filled with ample promises of profit. Indeed, the success he was to reap from these inventions was truly substantial, with the generous terms of a solid one-third partnership mentioned in the letter being only the beginning of the large financial return he was to receive. Bellamy agreed to add to his overseeing the production of the frames, cases, and shelves for his partners Eldridge Walcott and Jonas C. Young for $70 a week,[36] an excellent pay rate for the time, which must have been all the more pleasing compared to the uncertainties of working for the Navy on a project-by-project basis during an era of downsizing. Bellamy was to also earn $1,000 in November 1868 selling his patents for Masonic frames to a Henry C. Merriam of Charlestown.[37] In a gesture that reveals again how considerate and responsible he was toward the welfare of his parents, Bellamy made certain in May 1870 to negotiate a 5 percent royalty fee for his "designs of Masonic odd fellow and papal emblems to be applied to clock fronts frames hand glasses brackets and other articles," with the royalty payments being paid expressly to his father, Charles G. Bellamy.[38]

Though many of the documents pertinent to this aspect of Bellamy's career date from the 1870s, we now know Bellamy was already producing and profiting from his Masonic frames by July 1866.[39] As to the identity of the "Free Mason" from the navy yard who Bellamy speaks of as wanting to be his agent, all indications point to him having been a gentleman by the name of David A. Titcomb. Having partnered up on July 30, 1866,[40] Titcomb and Bellamy were listed as being in business at 11 City Square, Charlestown, in 1868, on the first floor of a Masonic lodge.[41]

Fig. 4.20 Business card for Titcomb & Bellamy, ca. 1866–72. *WAF*.

Their business card, [Fig. 4.20] further reveals that Titcomb was maintaining an office at 31 Washington Street, Boston.

Being "equal owners of all stock and materials"[42] regarding the frame business, partnership stressed parity and fairness; their business contract stipulated that if Bellamy and Titcomb could not agree on a certain matter, they would turn to third-party arbitration. In classic legal fashion, the arbitrators would be three people, one chosen by Bellamy, one by Titcomb, and the third by the two arbitrators already selected by the partners.[43]

While Bellamy was the creative genius behind the designs and the man who oversaw their competent manufacture, Titcomb was his "lawful attorney" in matters regarding the selling, transfer, assigning, and setting over of the Masonic picture frames.[44] (Walcott and Young appear to have only been financial backers for the business venture.) Yet Titcomb appears to have labored mostly in the role of salesman. In a letter to his father in March 1868, Bellamy notes, "I have been very busy since you left preparing a set of samples for Mr. Titcomb to take West."[45] Titcomb appears to have successfully sold Bellamy's creations in at least a dozen states by 1868: Vermont, New York, New Jersey, West Virginia, Virginia, North and South Carolina, Georgia, Alabama, Florida, Mississippi, and Tennessee.[46]

Fig. 4.21 John Haley Bellamy, E. N. Welch Clock Co., *Shelf clock with Independent Order of Odd Fellows symbols*: walnut, brass, ca. 1870. [20 x 14 x 3½ in.] *SRMM / DB*.

When not producing clocks, frames, and shelves for the Masons, Bellamy turned his talents toward producing works that appealed to other markets. From the credit reports compiled by the firm of R.G. Dun & Company (today's Dun & Bradstreet), we now know that he was busily setting up another business at 89 Court Street, Charlestown, in February 1869 that was independent of Titcomb, one that was founded upon the manufacture of clocks and jewelry. Within these credit reports we find a sober, objective window into Bellamy's work ethic and job performance, as well as the state of his business affairs at this moment in his career:

> J.H. Bellamy & Co. Clocks & Jewelry, etc.
> 89 Court Street
>
> Feby 23/[18]69. B[ellamy] is a Carver by trade is steady and Indus[trious] with small Means he owns some patents wh[ich] he considers valuable he lives in Charlestown Mass & is now doing bus[iness] there & only has a

Fig. 4.22. Photograph of John Haley Bellamy, *Eagle frame*: unknown wood, ca. 1866–72. *WAF*.

> stopping at 89 Court St. he has no pr[operty] at present but expects to Make arrangements to have one soon, has no established Cr[edit].[47]

Other artistic offerings produced by the (apparently short-lived) J. H. Bellamy & Co. include clock cases for other fraternal organizations like the Independent Order of Odd Fellows, [Fig. 4.21] and the Grand Army of the Republic and the Knights of Columbus.[48] One intriguing creation that may hail from the J. H. Bellamy & Co. workshop is the "Papal Clock," a design that was encrusted with the emblems and heraldic devices of the Catholic Papacy in a fashion that seems little different from his Masonic pieces. Designed to court the interest of America's growing Catholic populace, no examples of this clock are known to have survived, leaving nothing more than a handbill to gain an idea of their appearance:

> The wood-work of the Clock is carved, and the Pontifical emblems are presented in relief, as follows: Over the clock dial is seen the Pope's "Tiara," (or triple crown,) beneath which appears the "Keys of St. Peter." On one side of the dial is seen the "Crozier," (or Shepherd's Staff,) and on the other the "Mitre." [*sic*] In the centre, below the dial, is a correct photograph of Pope Pius the IX, on one side of which is the Cross, and on the other side the Open Bible, marked Matt. 16th Chapt. 18th and 19th verses; comprising, as will be seen, all the Pontifical emblems, and enclosed in ornamental scroll-work.
>
> The Clocks are suitable for mantel-shelves.[49]

Another of Bellamy's creations from this era was not concerned with secret societies and religious denominations but an appeal to patriotic Americans far and wide. With its strong classical lines and refreshing austerity, especially when compared to his Masonic designs, Bellamy's eagle frame [Fig. 4.22] was a marriage of solid classical proportions and streamlined minimalism. While not a novel idea (Britain's Isaac Ware had envisioned an eagle adorning a mirror frame as early as 1756[50]) Bellamy's eagle frame nevertheless represents a departure from the baroque styles of the Victorian era, a more accomplished achievement than the unrefined eagle-adorned furnishings produced in earlier eras. It was a herald of those inspired Bellamy designs soon to come.

BY THE EARLY 1870S, JOHN HALEY BELLAMY WAS THE driving force behind a successful emblematic frame and bracket business, one that granted him the funds necessary to keep himself afloat financially amid the uncertain times of the post-Civil War downturn in US naval construction and repair. With his work circulating throughout several states in the American South and Northeast, he had come a long way from the confusion and malaise of a love gone bad and his days as a figure and ornamental carver of limited reputation in Portsmouth, Boston, and Medford. It had not been an easy achievement: A fickle market and the whims of the US Congress and Navy had all conspired to make his vocation one fraught with periods of unemployment, forcing him to move often and become ever more resourceful. Keeping true to his calling, Bellamy had struck upon a novel market for his inspired designs and had thrived in the process. To the outside observer, John Haley Bellamy had arrived, yet his greatest triumphs were about to begin for the woodcarver from Kittery Point.

The Lancaster Eagle Part 5

"WORK ON THE BIG EAGLE WHICH IS TO BE PLACED ON THE BOW OF THE U.S.S. *LANCASTER* HAS BEEN COMMENCED IN THE MOULD LOFT AT THE NAVY YARD, BY THAT EXPERT CARVER MR. JOHN H. BELLAMY."[1]

With this one simple sentence, the *Portsmouth Times* announced to the world that construction had begun on "one of the largest and most elaborate"[2] figureheads ever to grace a US naval warship. Given how revered the *Lancaster* figurehead has become over time the lack of coverage seems disappointing, even anticlimactic. Yet Moses Victor Safford in 1941confirmed this was indeed had been the case, as he remembered the event. "I cannot recall that the construction and mounting of an eagle figurehead on the *Lancaster* was treated as a matter of any great interest at the Navy Yard" (MVS: *JHB*, Pg. 15).

As with so many other aspects of John Haley Bellamy's life, Safford serves as the prime window for understanding this segment of the great woodcarver's life. In fact, in regard to the construction and installation of the *Lancaster* Eagle Safford proves to be a particularly invaluable resource, one privy to the figurehead's development in intimate detail from inception to completion. As he would explain in later years:

> I was then a boy thirteen or fourteen years old, living in Kittery and crossing the Navy Yard twice and perhaps four times a day to take the Navy Yard ferry to and from school in Portsmouth . . . My daily path to and from school took me directly under the bow of the Lancaster while she was being rebuilt . . . I saw it grow under Bellamy's hand in the second story of the old granite building alongside the present graving dock . . . (MVS: *JHB*, Pgs. 11, 13)

With his continuous and intimate exposure to the figurehead throughout its evolution, Safford is more than qualified to recount how the *Lancaster* Eagle was created. We learn from Safford that Bellamy's work space at the Portsmouth Navy Yard on the second floor of an "old granite building . . ." was set in "a partitioned off space with windows overlooking the present site of the graving dock which was then a navigable channel running East and West between the old navy yard and Seavey's Island, the identity of which is now lost" (MVS: *JHB*, Pgs. 13–15). From this we can locate without question the building where the *Lancaster* Eagle was carved, a former sail loft built in 1837 known today as Building 7, a structure that is still standing and in use by the United States Navy. [Fig. 5.1]. "It was here and in the loft outside" Safford recalled, "that I as a boy just entering my teens saw the Lancaster's eagle figurehead evolve out of a collection of blocks of wood of various shapes and sizes" (MVS: *Eagle*, Pg. 12) Moreover:

> It seems to me that the eagle figurehead was put up to Bellamy like any other carving job and nobody doubted his ability to make good on it. Such interest in it which I recall as having been manifested by others was comprised in watching to see how Bellamy would meet the practical construction problems presented by the size

Fig. 5.1 Building 7, Portsmouth Navy Yard. This former mast house and sail loft hosted John Haley Bellamy's woodcarving shop during the nearly three decades he spent as an employee of the Portsmouth Yard after the American Civil War; it would witness the crafting of the USS *Lancaster* Eagle figurehead. *UNH.*

> and unwieldly [*sic*] shape of the proposed figurehead. I too was thus interested. I recall many sectional drawings on large sheets of brown paper and some "laying out" on the floor of the mould loft, and some use of "forms" as in the construction of a ship. Wooden blocks were built up somewhere else in the Navy Yard, approximately the size and shape as that portion of the eagle into which they were intended to evolve. These blocks were not too large to be taken into Bellamy's regular work room and carved down by him to fit properly into the completed eagle. To what extent the entire eagle was assembled in and the sections fitted together in the mould loft or somewhere else before the sections were finally joined together on the bow of the ship I do not know (MVS: *JHB*, Pg. 15).

These words agree with what we know today about the tools, procedures, and tricks nineteenth-century American ship carvers resorted to when fashioning a figurehead.[3] Combining these recollections with our present-day knowledge of figurehead construction, we can easily surmise that Bellamy took into ample consideration the shape of the *Lancaster's* bow and the rake of its bowsprit when envisioning the final creation. This idea is strengthened by Safford's claim that Bellamy "knew that anything ornamental on the bow of a ship could not be permitted to interfere with necessary features of the bow, with the head, or with the customary staying of the bowsprit and jib-boom" (MVS: *Eagle*, Pg. 11).

After considering the architectural specifics of the job and consulting with US Naval Constructor T. D. Wilson, Bellamy then began to sketch out the figurehead's design on heavy brown paper. Once a workable design was drafted, tradition bade that he next select the wood from which he would carve his mighty bird. Bellamy elected to use pine for the eagle, a cheap, plentiful material in nineteenth-century American shipyards and the overwhelming choice among Yankee woodcarvers for such work. Given its projected enormous size and that no single block of pine existed large enough to carve the eagle in a single piece, Bellamy was compelled to fashion the mighty raptor in sections, hewn from several blocks of pine. Yet rather than using just any pine blocks that happened to be available, he personally selected the wood to be used, keeping a sharp eye upon its grain and overall fitness for the project at hand. And then, with only mallets, gauges, chisels, and sandpaper, the carving began.

Assuming Bellamy approached this figurehead as the tradition dictated, he first "roughed out" the blocks on the floor, making preliminary cuts that would determine their basic shape and form. Once done, he then would hoist the blocks up on "horses," rope-and-pulley cradles secured to the roof of the workshop, to afford him the ease of access needed to carve the finer details of a large three-dimensional subject from multiple angles. To avoid splintering, Bellamy carved with the grain of the pine, adjusting the details he envisioned for the piece accordingly to its fickle

Fig. 5.3 Gleason's Pictorial, *United States Navy Yard, Portsmouth*: lithograph, July 23, 1853. *DP*.

CHAPTER 5

The Return to Kittery Point

JOHN HALEY BELLAMY'S TIME IN THE BOSTON area, working intermittently at the Charlestown Navy Yard, running his own woodcarving shop with his brother Elisha and peddling a host of inventions, was both artistically and monetarily successful. Yet by 1872, Bellamy had packed up his shop and returned home to the Piscataqua area, leaving Charlestown and a productive business in his wake, never to return again. On the surface this appears a perplexing and counterintuitive move that defies logic and has led past Bellamy biographers to draw an inaccurate assessment of the man and his reasons for relocating back home:

> Despite the apparent success of his business, John at about age thirty, began to develop a pattern of living that would prevent him from ever becoming financially secure. He was restless and never stayed at anything for very long. He was with Beecher just a few months. He studied at New Hampton Institute for only one semester. His association with Titcomb, though profitable, was of short duration.[1]

This simplistic vision of Bellamy as "restless" and erratic, skipping from situation to situation with no thought to his future, is another unfortunate misconception that has been handed down over time. Bellamy's brief stay in Laban Beecher's former shop was undoubtedly due to the economic crash of 1857 rather than any capricious attitude on the part of the artist. We now know he was enrolled for more than one semester at the New Hampton Literary and Biblical Institution, and that his partnership with David A Titcomb, spanning a period of six full years, while not lifelong was anything but brief. It is clear that the assumed "pattern" of fickle behavior in his life was nonexistent. Moreover, Bellamy's decision to return home was anything but an act of restlessness.

For a true and accurate appreciation of just why Bellamy came back to Kittery Point in 1872, diverse factors such as his family situation, the employment climate at American navy yards at that time, and the nature of the woodcarver's trade itself must be considered.

First, about the time of his return to Kittery Point, John Haley Bellamy found himself confronted with mounting familial duties. Born in 1807, Bellamy's mother was well into her senior years by the early 1870s.[2] Though she was of hearty Yankee stock and ultimately reached her eighties, it is reasonable to assume that time was catching up to Fanny Bellamy, forcing her unmarried eldest son to come home and be of assistance to her. Bellamy also appears to have had concern for his aging father, repeatedly offering him employment at his Kittery Point shop when Charles Gerrish Bellamy was still at work at the Charlestown Navy Yard. Interest in his parents' financial welfare may have also played a role in his decision to move back home, as Charles Gerrish Bellamy went bankrupt by the age of sixty.[3] (Charles Gerrish Bellamy, born in the year 1811,[4] had thus entered into bankruptcy by 1871, almost exactly when his eldest son made his return home.)

That the Bellamy family patriarch had indeed fallen upon hard financial times by this point in his life is suggested by his working as a boat builder in Charlestown rather than running for public office and entertaining political allies in high style in Kittery as he had done in decades past.[5] Documents of the late 1860s and early 1870s show John was providing abundant victuals and supplies for his parents at his own cost. Furthermore, along with his brothers Elisha and Charles, in 1869 he purchased homeowners' insurance on the family property and included their mother's name on the policy.[6] He resided at home specifically to "let mother have the benefit of what I pay out here" in terms of room and board.[7] How great or small a role these familial factors played in drawing Bellamy back home is unknown, yet the evidence clearly shows they influenced his decision to a significant degree.

Second, it is imperative to remember that woodworkers in nineteenth-century America were a somewhat itinerant class of laborers, even when they had gainful employment. House carpenters and shop-bound furniture makers were beholden to the whims of the market and dramatically subject to the effects of industrialization in their professions. Productivity at commercial shipyards was governed by fluctuations in the market as well, while government yards were obligated to the budgetary dictates of Congress. The ultimate outcome was that American woodcarvers more often than not found themselves moving from job to job, and even town to town, out of necessity rather than restiveness, turning their hands to whatever tasks they could secure.

This explains why Bellamy moved about during the late 1850s and 1860s, setting up shop in Portsmouth and toiling away in Boston, Medford, and Charlestown. In retrospect, Bellamy was more sedentary than many fellow woodcarvers of his time. Had he practiced his trade elsewhere, he may well have experienced many more changes of address throughout his life. Thus, for Bellamy to leave Charlestown in 1872 and pursue his craft elsewhere was not unusual—in fact, just another occasion where his work dictated where he should live.

The transient nature of the nineteenth-century woodworker's calling provides us with the most concrete of reasons for Bellamy's homecoming, especially after the dramatic contraction of the US Navy's warship fleet at the conclusion of the Civil War. Layoffs, or "discharges," happened with frightening regularity at the Charlestown Navy Yard throughout the post–Civil War era, making it an unreliable source for income at best. Yet back home in his native Kittery was another navy yard, where he had already worked during the war. While undergoing its own postwar production problems, it may have offered Bellamy a distinct ray of hope.

Like the Charlestown Yard, the Portsmouth Navy Yard [Fig. 5.3] entered into an era of high productivity during the Civil War and the postwar environment brought this era of industry to a shuddering stop. [8] Despite this discouraging circumstance, the Portsmouth Navy Yard provided Bellamy with one thing the Charlestown Yard could not: gainful employment as an ornamental woodcarver.

The date he was taken on again by the Portsmouth Navy Yard was probably in September 1873, as alluded to in a letter from Fanny Bellamy to her husband Charles dated September 25, 1873: "John left Cutts job and has gone on to the Navy Yard carving. [He] will have a year's job if he keeps straight, and good pay."[9] A sterling reputation with key individuals at the Portsmouth yard seems to have assured him the regular employment that he could not find back in the Charlestown facility. Bellamy's *Portsmouth Herald* 1914 obituary states Bellamy's "services were highly valued by the late Naval Constructors Wilson and Hitchborn [*sic*],"[10] two individuals intimately involved in the Yard's daily operations. Bellamy himself writing his father in November 1873 commented on why his employers so valued him and held him in high esteem. "Well I am still there and have the assurance that I can stay as long as I desire. They have had much trouble on their last ships in having the work come right and believe that I understand my business they seem well pleased."[11]

With the trust and good opinion of Wilson, Hichborn, and others assigned to determine who was to be hired for the various projects that trickled into the Portsmouth yard, Bellamy was thus assured first pick of whatever ornamental woodworking projects needed doing, being "employed frequently for long intervals" at that facility (MVS: *Eagle*, Pg. 10). "For a long period of years, beginning prior to the Civil War and continuing until after the opening of the present century, he carved figureheads, ornamental work for sterns, panels for gangways, and sundry other decorations for other parts of naval and mercantile craft, large and small."[12] Aside from his efforts concerning merchant vessels and yachts, Bellamy was also remembered as having been employed in "carving scrolls for the sterns and bows of . . . naval ships" (MVS: *JHB*, Pg. 2). These words are confirmed by Bellamy's obituary. "He began his work as an expert wood carver at the Portsmouth navy yard during the war of the Rebellion and he was employed there until 1900. While at the navy yard he made the figure heads and did other carving for the war vessels constructed during that period."[13]

From catheads and billetheads to sternboards and figureheads, no facet of a warship's exterior that could accommodate decorative woodcarving appears to have been beyond the reach of Bellamy's hand. However, aside from his eagle figurehead for the USS *Lancaster* and a sternboard he crafted for the USS *Wabash*, the only other documented pieces of exterior woodworking Bellamy executed for the US Navy that we can still view today are gangway boards. Moses Victor Safford held particularly vivid memories of these carvings:

> What appealed to me especially as a boy when visiting his workshop at the Navy Yard were the elaborate carvings on the mahogany panels which formed the sides of the gangways in the bulwarks of the old wooden Navy ships. I understood that Bellamy carved them to be shipped to other Yards. I have often wondered what became of them when the old wooden ships were broken up. (MVS: *Eagle*, Pg. 10)

Exceedingly rare, today only four are known to still exist. Indeed, all four gangway boards appear to hail from the bark-rigged screw sloop-of-war USS *Enterprise*, [Fig. 5.4] a vessel that, along with her sister ship the USS *Essex*, was the last wooden warship ever to be constructed at the Portsmouth Navy Yard.[14] Launched in 1874, the *Enterprise* gangway fittingly enough sported "an artistic adaptation of the Arms of the United States" (MVS: *JHB*, Pg. 2), with two large boards featuring, respectively, an eagle grasping the top of an anchor tightly within its talons, and an American shield with flags and a splash of stars radiating from behind [Fig. 5.5], and several smaller repetitive panels sporting the American eagle resplendently perched atop a shield [Figs. 5.6 and 5.7]. While the two larger, more elaborate boards would have been installed at either the top or bottom of the gangway, the two smaller pieces, with their recurring design, would have been mounted anywhere along its length.

From the individual feathers upon each bird to such minor details as the definition marks running along the top rim of the shield and painted flags behind it, it is readily apparent that the two smaller pieces are nearly identical. This may lead one to conclude that these pieces may have been produced with the help of apprentices and assistants employing paper patterns drawn by Bellamy, as he had done with his Masonic brackets, clocks, and shelves. Yet on close examination it becomes evident that these two particular pieces were carved by hand by Bellamy and Bellamy alone; subtle differences between the two eagle heads verify the fact that they were indeed hand carved rather than machine cut. With their exquisite detail and robust, powerful presence, and from the forceful hand with which they were carved, these two mahogany boards stand as superlative examples of Bellamy's work for the United States Navy, revealing the high level of workmanship Bellamy placed into even the most routine, everyday types of carving he was called upon by the Navy to execute.

However, the same cannot be said for the two larger boards. Though both bear the distinctive characteristics of a Bellamy design, only one—the curvilinear shield with the field of stars behind it—appears to have been executed by his accomplished hand. Upon close inspection, the board featuring an eagle atop an anchor shows itself to be of an inferior quality to Bellamy's usual handiwork, especially in regard to the lack of definition in the bird's feathers [Fig. 5.8], a glaring disparity when compared

Fig. 5.4 Wooden screw sloop-of-war USS *Enterprise*: photograph, 1874. *UNH.*

to the two smaller gangway boards from the same ship mentioned above. Why does this disparity exist? Is this particular gangway board the handiwork of a fellow woodcarver at the Portsmouth Navy Yard, crafting a design originally conceived by Bellamy, following the master woodcarver's outline to the letter? Could Bellamy, rushed for time, have turned to the use of assistants for navy yard jobs in a fashion little different from how he ran his frame and bracket business in Charlestown? Or is this piece a crude copy of an original Bellamy design, crafted at another time by another artist completely independent of him, perhaps meant to replace the original (perhaps damaged) Bellamy gangway board? At present each of these theories holds equal weight, teasing us with their possibilities and begging even more questions about how the master woodcarver ran his studio at the Portsmouth Navy Yard.

For all the carvings he was called upon to craft for the exteriors of US naval vessels, it appears that the greater number of Bellamy carvings was actually to be found within their interiors. As it has been told:

> The distinguished foreign visitor who boarded an American warship in a foreign port sixty or seventy years ago saw . . . Over the low entrance to the old time cabin there probably would be seen a typical Bellamy "wall eagle" with a carved draping of the American flag, or bearing in its beak a carved pennant on which appeared some motto, or a star, or stars, or the colors of the flag.

Fig. 5.5 John Haley Bellamy (with others?), *USS* Enterprise *gangway boards*: mahogany, 1874. [measurements unknown] *PA.*

Fig. 5.6 John Haley Bellamy, *USS* Enterprise *gangway board*: mahogany, 1874. [35¼ x 21½ x 2 in.] *Private collection, AKA.*

Fig. 5.7 John Haley Bellamy, *USS* Enterprise *gangway board*: mahogany, 1874. [35¼ x 21½ x 2 in.] *Private collection, AKA.*

> Perhaps the arms of the Captain's chair as he sat at dinner terminated in eagles' heads. (MVS: *JHB*, Pg. 2)

That Bellamy produced interior woodcarvings for the Navy comes as little surprise. His mentor Samuel Dockum had engaged in such work for commercial ship owners during the time Bellamy was apprenticed to him. It stands to reason that Dockum would have included him in the task of appointing the cabins of vessels, such as the clipper ship *Josephine*, and in the process by one degree or another, imparted onto him the skills necessary to execute similar work in the future. What is more, Bellamy was well skilled in crafting interior decorative woodwork for homeowners in the Kittery-Portsmouth area, a legacy gained from his father and one which shall be discussed in greater detail in chapter seven. Yet while examples of Bellamy's exterior woodwork for US naval craft are exceedingly rare and hard to find, his interior pieces are entirely nonexistent. To date, no example of interior woodwork crafted for a US Navy vessel by Bellamy has been definitively identified. Tragically, beyond the previously quoted description we have little to guide us in envisioning the rich appointments of a captain's cabin decorated by Bellamy.

Perhaps the best tool for helping us imagine just how a cabin or wardroom appeared after having received Bellamy's ministrations can be found in a series of photographs of the USS *Vandalia*, shot in 1886 while docked in New York and 1887, when on duty in the Pacific. While there is no evidence to suggest that Bellamy actually ever worked upon this particular warship, the *Vandalia's* captain's cabin, [Fig. 5.9] after cabin, [Fig. 5.10] and wardroom reveal the surprisingly tasteful woodwork that was typically found in the cabins and wardrooms of American warships during the latter nineteenth century. They offer us a glimpse of how such a cabin might have looked after having received Bellamy's inspired artistry.

The amount of artistic output Bellamy achieved during his years with the U. S. Navy remains unknown. With almost nonexistent records and only a few stray surviving examples to view, it is impossible to gain an accurate appreciation of just how varied and diverse his government commissions were. Yet, it is evident that Bellamy's creations had an eager audience within the federal government, with his carvings traveling far beyond the ships that pulled in and out of the Portsmouth yard. According to Safford, "A great deal of the work which he did at the Kittery [Portsmouth] Navy Yard was shipped away to other Navy Yards and I believe that some of it went to Government departments other than the Navy" (MVS: *Eagle*, Pg. 9).

Working at the Portsmouth yard agreed with the woodcarver, despite his being paid the rather typical Yard wage of \$2.32 a day.[15] In a letter written soon after returning to the

Fig. 5.8 John Haley Bellamy (with others?), Detail of eagle, *USS* Enterprise *gangway board*: mahogany, 1874. *PA*.

Yard in 1873 he assures his father, "I am comfortable," adding, "I have a nice comfortable shop—The pattern shop over in Construction . . ."[16] Regarding this shop, Bellamy's nephew concurred, stating that the Navy had given Bellamy a "commodious work room." Safford also tells us that his famous uncle enjoyed

Fig. 5.9 Captain's cabin, USS *Vandalia*: photograph, 1886. *NHHC*. The gentleman featured in the photograph is Captain Henry L. Howison, the *Vandalia's* commanding officer.

Fig. 5.10 After cabin, USS *Vandalia*: photograph, 1886. *NHHC*.

a high degree of autonomy within his shop, where he "apparently worked without supervision, as 'his own boss'. Any check on his time seemed unnecessary for he was always busy when on a job. He worked as though trying to finish it" (MVS: *JHB*, Pg. 13). The high esteem in which Bellamy was held by his navy yard employers stands as much a credit to his artistic skills and knowledge of naval architecture as to his work ethic, conferring onto him an eminence beyond that of his fellow yard workers.

"Everybody understood Bellamy's special status and why he had it," Safford claimed, adding, "Bellamy was 'naval minded' as indicated by his treatment of the emblems of the United States Government and naval officers with many of whom he maintained friendly personal relations probably tended to appreciate his art better than the civilians in authority" (MVS: *Eagle*, Pg. 11). Among those naval officers with whom Bellamy enjoyed a particularly "long friendship" (MVS: *JHB*, Pg. 1) was Commander Merry, to whom Bellamy would go so far as to present an "ambitious piece of wood sculpture," a life-size eagle with spread wings, which was widely considered by those who saw it to be "one of the artist's best achievements."[17] This "special status" Bellamy enjoyed with the Portsmouth yard's senior officers and foremen appears to have carried him away at times, with the "essentially courteous, quiet mannered, and good natured" (MVS: *Eagle*, Pg. 5–6) artist becoming a touch haughty.

> One memorable occasion where his highhanded sense of self-importance got the better of him involved his having been sent to repair the sternboard of a warship that had recently arrived at the Portsmouth Navy Yard. Reporting for duty at the unnamed vessel, the famed woodcarver found—much to his chagrin—that he had only been equipped with a narrow boatswain's chair suspended from the vessel's spanker boom by a slender rope and tackle. Determining that the only outcome working atop this frail contraption would involve a fall into the harbor below, Bellamy immediately demanded an interview with the ship's captain. "I am Mr. Bellamy," he is recorded as having said upon gaining an audience with the captain. "I have been sent to repair some carving on the stern of your ship. I find that a boatswain's chair has been provided for my use in doing this work. I wanted you to *see* me, because you evidently thought that I was some sort of a damned monkey that

> could work with his hands and hang on by his tail. If you want me to make the repairs called for you will have to order a suitable platform."[18]

As the narrator of this incident remembered it, "The specification was promptly met."[19]

This near-comical exchange between the artist and the anonymous captain stands as a rare example of the Bellamy who would "not allow anyone to impose on him," and who "could be imperious in his treatment of others without much provocation" (MVS: *Eagle*, Pg. 5–6). Yet it also seems to have given rise to a host of fantastic stories that portray the woodcarver as a volatile brawler during his time at the Portsmouth yard, one who was both cantankerous and unpredictable. Fortunately, Moses Victor Safford sought to address such stories:

> In his lifetime Bellamy was widely known as an interesting character. He was always doing something or saying something unexpected. Naturally anecdotes were told of him . . . which are entirely untrue. Recent numerous articles and even books about old wood carvers and accounts of the rescue of the *Lancaster's* eagle from a junk shop have produced a new crop of stories about Bellamy which I must regard as pure fiction. I have read accounts of quarrels with naval officers regarding the design, construction and the placing of this eagle on the *Lancaster* and of Bellamy's refusal to go on with the work if he could not have his own way. He knew as well as anybody ship construction and the practical considerations to be taken into account in attaching anything to a ship anywhere, and would be very unlikely to make unreasonable demands for the sake of artistic effect. He was not dependent on his Navy Yard employment for his living and would undoubtedly have quit if told to do anything which would have destroyed the artistic character of his work, but I do not believe that anything ever happened to oblige him to take this stand. In all his employment by the Navy I myself never heard of his having any trouble with anybody connected with the Navy in a position of authority or responsibility (MVS: *JHB*, Pg. 16).

One tale in particular, that of Bellamy assaulting a naval officer with one of his carving tools, received particular attention from the artist's nephew/biographer:

> I recently read a story about his chasing with a carving tool a young Naval officer who woke him up while taking a nap on his uncompleted *Lancaster* eagle. None of the sections of the eagle before they were put together at the *Lancaster's* bow were big enough for a man to lie down upon, so that part of the story is certainly an invention. Bellamy never loafed at his work. When sick he never hesitated to go home and stay there . . . He was not the kind of a man who would chase anybody with a dangerous weapon even as "horseplay".[20]

Safford was right to be suspicious of this story. In the latter decades of the nineteenth century, Navy Yard employees generally lost their jobs for one of two reasons: reduction of workforce necessitated by a contraction in work or shortage of funds, or because the individual merited discharge due to misbehavior.[21] Such misbehavior could comprise anything from the seemingly benign offenses of "false muster," and "insulting the watchman," to more serious crimes, including "stealing," "idleness during working hours," "fighting," and, most tellingly, "drunkenness and sleeping during working hours."[22] Regardless of the respect and friendship he enjoyed among the Yard's senior officers and civilian foremen, it is hard to imagine Bellamy not being fired (if not arrested) for the double offense of being asleep at his job and physically assaulting an officer in the United States Navy with a dangerous weapon. When appraising Bellamy's "character of conduct and work," his Navy Yard supervisors were moved to describe both as being "excellent," an unlikely evaluation to say the least if he was known for brawling and sleeping on the job.[23] And lastly, the actual publication responsible for proffering this implausible story loses further credibility in its mistaken claim that Bellamy constructed the *Lancaster* Eagle figurehead in the

Fig. 5.11 C. W. Wright, *View of the Interior of the United States Navy Yard, Charlestown, Mass.*: lithograph, ca. 1853. *NPS, Charlestown Navy Yard, BOSTS 8635.*

Charlestown Navy Yard [Fig. 5.11], thus consigning this tale to the realm of the legendary.[24]

When not busy at his job in the Portsmouth yard Bellamy was apt to be found spending his days at the Pepperrell Mansion, surrounded by several of his kith and kin. Though Charles Gerrish Bellamy was off in Charlestown he kept in regular contact with his family, primarily through his eldest son, and it is through this exchange of touchingly personal letters that we gain an intimate glimpse of how daily life unfolded in and around the Bellamy household. Oftentimes more than one member of the family would contribute a paragraph or two to these letters, their collective words keeping the Bellamy family patriarch abreast of all the mundane yet precious everyday comings and goings of his family during his absence. In a letter from September 21, 1873, John Haley Bellamy paints for his father (and us) a particularly vivid portrait of the Bellamy clan spending a typical, uneventful Sunday under the roof of the Pepperrell Mansion.

> Dear Father.
>
> It being Sunday we are all at home—Elisha from the Yard Charles & I from the foreside—Well after brushing around as usual Sunday morning we all took breakfast as usual Sophronia and little Jack included. Then Sophronia prepared a boild [*sic*] dinner saying now Elisha you go out with me and we will cut out some of the loose heads of cabbage for you know Mother saves the full heads for winter. So out they go and after much hunting find such as they want (for we are to have the largest lot of good cabbage this season of any season since I can remember) as for beats [*sic*] Elisha says there will be 40 bushels—I don't think we will have that number but we shall have quite a lot—some of them measure over six inches and they are all very large we also have a lot of beans corn &c&c[25]

Interestingly, Bellamy would conclude this letter by making an announcement to his father, one which concerned a new business venture he was pondering.

> . . . I am now making arrangements to have a little business this winter in the shop and if I can get together capital enough before the snow blows—I shall be able to have work enough for all of us in case business outside should fail us—I know what to do and how to do it. So shall try to mature it to my own liking. Remember me to Mother and Laura—Josephine & Oliver. Write when convenient and wrish [*sic*] easy . . .
>
> Yours,
>
> J. H. Bellamy[26]

Ensconced back home within the Pepperrell Mansion, surrounded by his family and producing incredible carvings for the United States Navy, John Haley Bellamy was now prepared to embark upon a new chapter in his life, one which would come to both forever define him as an artist and bring his handiwork into American households far and wide.

THE LANCASTER EAGLE PART 6

THE CRAFTING OF THE *LANCASTER* FIGUREHEAD STRETCHED FROM DECEMBER 1879 TO SOMETIME BEFORE AUGUST 21, 1881, WHEN THE *LANCASTER* OFFICIALLY WENT BACK INTO COMMISSION.[1]

The end result was a titanic monolith of wood sheathed in gleaming gold leaf, ten feet high and with a wingspan of eighteen feet eight inches [Fig. 6.1]. Assembled piecemeal on the bow of its new home and bolted into place, its sweeping wings reinforced with stout wrought iron straps, the installation of the 3,200-pounds-plus figurehead was a memorable sight for those who were fortunate enough to witness it:

> . . . the eagle was put on a platform at the bow of the Lancaster in sections which could be conveniently handled, and I remember that on our way to and from school for several days, at least— perhaps for a week— my schoolmates and I stopped and watched Bellamy on this platform supervising the bolting of the sections of the eagle together and the whole affair onto the bow of the ship (MVS: *JHB*, Pg. 15).

While today this figurehead is recognized as "one of the most striking objects in the body of American decorative art,"[2] when it made its debut on the *Lancaster's* bow, the event merited no special attention. Incredibly, there was no dedication ceremony to commemorate the figurehead after it was installed, nor was there a launching ceremony for the rebuilt, refurbished warship it now sailed upon (MVS: *JHB*, Pg. 17). Local papers failed to make mention of the figurehead when the ship finally departed the harbor, issuing nothing more than the blunt statement, "The U. S. S. *Lancaster* steamed out of the lower harbor in good style, Monday morning, on her way to New York."[3] Moses Victor Safford offered an explanation why the figurehead failed to illicit celebration and comment.

> it . . . was not a very conspicuous affair notwithstanding its gilded outstretched wings. It was overshadowed by the bowsprit and the head and was obscured when viewed some distance away from the water to some extent by bob stays, martingale, back ropes, etc. The wide spread net which these old naval ships carried to catch the head sails when lowered also tended to detract from the effectiveness of any figurehead (MVS: *Eagle*, Pg. 13–14).

A painting of the USS *Lancaster* by an unidentified Chinese artist around 1882 readily confirms Safford's words [Fig. 6.2]. For despite the dramatic contrast of its gilded feathers against the black hull of the *Lancaster* and its enormous bulk, the figurehead was ultimately dwarfed by the warship. Few, if any, figureheads ever avoided this problem. No matter how they were painted or gilded and despite the attempts of woodcarvers to imbue their creations with an august presence, figureheads were doomed to a life of general anonymity beneath the bowsprits of the vessels they served, destined to elicit notice and comment only when in port, where people could draw near and view them up close.

The true artistry of Bellamy's creation resides not in its sheer size or the feat of engineering it took to create, assemble, and install it but rather in its aesthetic qualities. Safford was right to note, "Bellamy designed his eagle figurehead in size, in form and by the angle of its wings to blend with bowsprit, bob stays, back ropes, nettings and martingale as the head gear fitted harmoniously into the scheme and outline of masts, yards and stays" (MVS: *JHB*, Pg. 19). Never meant to overshadow the ship upon which it soared, Bellamy intended the *Lancaster* Eagle to seem a living, organic outgrowth of the vessel itself, inseparably entwined with its structure and identity. [Figs. 6.2 and 6.3]. In so doing, Bellamy had also created what critics have cited as "a turning point in the visualization of the American eagle,"[4] one that departed from the ponderous realism of previous generations of eagle carving artisans and introduced to the world a simpler, more streamlined, yet infinitely more expressive form of sculpture. Bellamy had done more than craft a convincingly realistic sculpture—majestic, graceful despite its weight, and filled with a raw power—with a surprisingly small amount of detail. He had also created a monument to the United States itself. The piece embodies the exceeding confidence of Bellamy's America. It is a prophetic piece of sculpture, reflecting the strength and steely determination of a country that within less than two decades would boldly soar forward into international affairs, determined to take its rightful place as a global power and world leader.

Fig. 6.1 John Haley Bellamy, *USS* Lancaster *Eagle figurehead*: painted and gilded pine, as seen today in the entrance hall of The Mariner's Museum, Newport News, VA. 1879–1881. [120 x 232 x 152 in.] *MM.*

Fig. 6.2 Unidentified Chinese artist, *The USS* Lancaster *in China*: oil on canvas, backed on masonite, ca. 1881. [16½ x 22 in.] *PMS / RM.*

Fig. 6.3 USS *Lancaster* Eagle figurehead in its final years of service: photograph, ca. 1921. *MM*.

CHAPTER 6

Glory and Majesty Captured in Whittled Pine, Part I: The Bellamy Eagle

IT IS NOW APPARENT THAT THE FAILURE OF J. H. Bellamy & Co. combined with changes within the operation of the Charlestown Navy Yard in the early 1870s, a more favorable employment climate for ornamental ship carvers at the Portsmouth Navy Yard, and a number of familial obligations back in Kittery all played their parts in drawing John Haley Bellamy back to his hometown. One of the earliest letters we have from him from the time of his homecoming is dated September 17, 1872. He had already been residing at the Pepperrell Mansion for a bit of time. Though he may or may not have already been working at the Portsmouth Navy Yard, he had launched a new business almost immediately upon his return to the area. With typical candor he notified his father, "All is well and moving on as usual. I am still in the shop making Eagles[;] have some very fine birds on hand."[1]

This passing mention to his "making Eagles" is the first distinct reference we have of Bellamy embarking upon a new and dramatic chapter in his career as a woodcarver, one which would soon come to define him, easily outshining whatever reputation he had already established with his Masonic works. The origins of this new facet to his oeuvre hearken back to his days in the Charlestown shop at 11 City Square. That the inventive Bellamy had eagles on his mind at that time is evident when viewing a correspondence to his father dated August 8, 1868. While the letter itself makes no mention of the national symbol, Bellamy saw fit to fashion the letter C in *Charlestown* at the top of the page into the proud head of an American eagle [Fig. 6.4]. An act of pure whimsy, it nevertheless betrays that he was toying with a subject matter not relevant to the Masonic, Odd Fellows, and Papal-themed creations he was then crafting. That Bellamy had not only been toying with this new motif but actually made a few examples in Charlestown is shown in a letter to his brother Charles of December 12, 1872. Bellamy, who was by then well-ensconced back in the Kittery/Portsmouth area, asks his brother in Charlestown to go to his old shop and "send me the patterns of the birds father and I made for Jones. They hang over the bench on a nail where I used to keep my saws."[2]

Of course, eagles were not a new subject for Bellamy. His work for the Navy would have demanded crafting heraldic eagles for the interiors and sterns of American warships. His surviving model of a clipper ship suggests Bellamy had fashioned at least one eagle figurehead for a commercial watercraft, and he had

Charlestown Saturday
Aug 8th
1868
Dear Father
I shall

Fig. 6.4 Personal correspondence from John Haley Bellamy to Charles Gerrish Bellamy, August 8, 1868. *WAF*.

already turned out at least two different frame designs sporting the national symbol for commercial consumption. It was not the eagle itself that was new for Bellamy but rather his plans for making it the very focal point of his career, a trademark design that would both define him as an artist and show the way toward a new form of representational art.

In choosing the eagle, Bellamy opened a large and willing audience for his work. His Masonic-themed labors cashed in on a popular movement of the hour, yet they were also hamstrung by a complete lack of interest among non-Masons. Regardless of how popular the Masonic movement was in America at that time, it was a finite audience. Yet the American eagle as a decorative motif enjoyed a massive and widespread appeal across the country, one not constrained by social class, education, gender, race, or selective membership in a secret society.

As a symbol the eagle was experiencing a pre-eminence it had not held since the Federal era, when the country was new and Americans were eager to express their nationalism. In the wake of the federal government's successful restoration of the Union, a resurgence of American pride and nationalism erupted. Bellamy, like so many of his countrymen, was living in a time of soaring expectations and national chauvinism, one marked by peace, prosperity, unprecedented national growth, and ensuing high optimism. Fortunes were to be made in steel mills, coal mines, and manufactories. Railroads crisscrossed the continent, uniting people on a scale never before imagined in human history. A gilded age of grand seaside cottages, massive hotels, and fanciful Victorian mansions was blossoming. The American eagle naturally represented the spirit of the hour, the prime symbol of American power, freedom, independence, and hope.

Victorian tourism made its debut in the Kittery area hand in hand with Bellamy's return in 1872 with the debut of hotels like the Pepperrell Hotel in Kittery Point and the Appledore House [Fig. 6.5] on the nearby Isles of Shoals.[3] What had once been a town solely focused upon shipbuilding and the sea increasingly turned its attention to meeting the needs of summer visitors in search of cool climes and pleasing diversions. A speedy rail service from Boston and a network of local trolleys and steamboats allowed affluent Bostonians and New Yorkers to make the pilgrimage northward. To accommodate these visitors grand hotels arose like gingerbread monoliths seemingly overnight. That John Haley Bellamy sought to exploit this new market in his hometown is clear from letter to his father in October 1872:

> I should have written you before this but have been quite busy on two eagles—which are for the Latons [*sic*] on the Isle of Shoals—one of them is to go on the big Hotel and the other on the dome of his gass[*sic*] house—they are both quite large—one of them stands from tip to tail to the back three ft from tip of the wings over eight feet—so you see it is quite a bird—stands on a ball—and I call it a nice bird—from the time I struck into the block until finished I was just three days for that I get $35 cash and for the other $15—with a prospect of two more. They will both be ready to deliver this week—so you see that they have not got eagles enough yet—I have another nice block in the shop for another as large as the largest. Shall make one to have on hand doubtful any risk in making it—the three blocks delivered at the shop door only cost me one dollar and forty cents—so you see that in three days I can make one of them worth more—shall give mother some cole [*sic*] out of it . . .
> Yours
>
> J. H. Bellamy[4]

By providing area hotels with his services, Bellamy not only created striking examples of a popular decorative motif but showcased his talents to a large throng of prospective clients hailing from several corners of the country. His decision to relocate to Kittery and pursue this new direction in woodcarving was immediately profitable as is evident in the generous sum he records as his payment for the Leighton commission and in the fact that other large, lucrative commissions immediately followed. Writing to his brother Charles in December, Bellamy not only asks him to send much-needed tools and patterns he had

Fig. 6.5 The Appledore House, Isles of Shoals: photograph, ca. 1900. Note the large eagle in the round by John Haley Bellamy perched prominently atop the hotel's cupola. *PA.*

left behind in the Charlestown shop, but incidentally informs today's audience that he had already secured investors and business partners "for the manufacture of eagles."

> I received your letter also Mother's. The vices came safe—I shall today sign articles of agreement with partners here for the manufacture of eagles—shall have a nice office and be boarded at the same hotel where I am now staying (The City Hotel.) Today I am to make drawings of various size birds—and then select the lumber and they will all be sawed at the mill—so you see I shall get rid of a large lot of work. I got the situation by carving a 4 foot eagle that stormy day Monday—and showing it to the partners. They are to have one hundred of that kind carved—and I don't know how many others . . . [5]

Fig. 6.6 Attributed to John Haley Bellamy, *Outdoor eagle executed in the round*: walnut, ca. 1872. [30 x 60 in.] *Private collection, CO.*

Eagles were fast becoming the very crux upon which his productivity as a woodcarver centered outside of his work for the Navy, with Bellamy supplying a number of public and private institutions with his raptors. Tragically, whether due to the ravages of time, the elements, or the occasional "Act of God" like the fiery conflagration that engulfed the Appledore House in 1914,[6] many of these seminal creations failed to survive. Yet some remarkable specimens have managed to endure into our modern day, offering us an understanding of the workmanship and stylistic traits of Bellamy's spirited early eagle designs.

One of the most uncommon Bellamy carvings ever discovered, an eagle executed in the round [Fig. 6.6], has come to light in recent years. It offers an idea of what his other early rooftop eagles may have looked like. Crafted out of black walnut, the

Fig. 6.7 John Haley Bellamy, *Large American eagle atop an out-of-round ball*: pine, ca. 1866–1872. [30¾ x 61 x 18 in.] *Private collection, AKA / DB.*

Fig. 6.8 Enlarged view of the large eagle executed in the round by John Haley Bellamy, situated atop the Appledore House, Isles of Shoals: photograph, ca. 1900. *PA*.

great raptor is captured at the exact moment of alighting atop its perch, delicately shifting and adjusting its weight while its wings begin to fold in as it prepares to roost. In this posture we see Bellamy executing a thoroughly complete and realistic depiction of his new subject matter of choice, not only in regards to its appearance but as concerns its behavior as well. For unlike so many of his later eagle carvings, this piece is not allegorical or symbolic in any way. Nor can the presence of stylized traits be detected in its bold, august presence. Within this sculpture Bellamy has done nothing less than offer the illusion of an actual living eagle having just landed. Capturing the nuances of such behavior so compellingly was a feat that surely required him to study living eagles in their natural habitat. For a pedestrian looking up at this carving from below, the effect would have been quite convincing, if not outright breathtaking. A triumph

Fig. 6.9 Detail of breast feathers and talons, *large American eagle atop an out-of-round ball*, pine, ca. 1872. *Private collection, AKA/ DB.*
The deep, thick cuts and incisions that define this sculpture's feathers lack the sensitive touch of Bellamy's later work. While it might be an early work by Bellamy, it seems more probable that this treatment of the feathers was intentional, serving to grant greater definition to the feathers and render them visible to viewers observing the sculpture at a distance many feet below. As well, note the overall lack of definition and scales to the talons done intentionally by Bellamy as a shortcut, as the talons could not be seen from below and thus did not necessitate laborious detail. Theories that the talons were executed by a less capable apprentice or that Bellamy had yet to master this particular aspect of eagle sculpture at the time (being new subject matter for him), cannot be discounted entirely.

Fig. 6.10 Signature of John H. Bellamy, inscribed into the right side of the out-of-round ball, *large American eagle atop an out-of-round ball*: pine, ca. 1872. *Private collection, AKA, DB.* Bellamy's decision to inscribe his name upon the bird's out-of-round ball maybe intended to ensure that admirers knew who its creator was at a time when the Bellamy name had yet to become synonymous with such creations.

in every respect, the piece stands as the most realistic of those rare three-dimensional Bellamy carvings known to still exist today, one that required a fine turn of the artist's imagination to conceive, let alone execute.

Another precious example of Bellamy's early three-dimensional work is a large American eagle atop an out-of-round ball

Fig. 6.11 Detail of *S*-shaped cuts along the back of wings, *large American eagle atop an out-of-round ball*: pine, ca. 1872. *Private collection, Courtesy AKA / DB.* The presence of these cuts covering the back of the wings of a bird carved fully in the round is encountered only on some examples of Bellamy's handiwork, usually in instances where the back of the sculpture would not be easily viewed due to its being displayed on high.

[Fig. 6.7]. Hewn from pine and originally stained to resemble mahogany, the three-dimensional properties and broad wingspan found within this piece also strongly suggest that it was meant to be seen from below, crafted specifically to crown the roofline of a Kittery/Portsmouth-area business. This hunch is strengthened when noting how this eagle's distinct profile bears more than a passing similarity to the eagle photographed atop the cupola of the Appledore House [Fig. 6.8], and goes far to explain the many peculiarities found within it [Figs. 6.9–6.11].

However, before declaring outright that these two eagles once loomed large over the Piscataqua region's skyline, it must be remembered that back in June 1859 Bellamy had made a point to advertise himself on his business cards as a carver of garden figures. (This was *exactly* the function one of Bellamy's large American eagle atop an out-of-round ball served as a focal point in the back yard of painter John Benson's Kittery Point summer home [Fig. 6.12].) Whether these two eagles once sat proudly on the crests of area hotels and department stores,

Fig. 6.12 Artist John Prentiss Benson applying a fresh coat of paint to the column supporting this *large American eagle atop an out-of-round ball* by John Haley Bellamy: photograph, ca. 1922. *PA.*

only to be pressed into serving as garden figures when their original homes had been remodeled or demolished, or were crafted as garden figures from the beginning, Bellamy's outdoor eagle executed in the round and his large American eagle atop an out-of-round ball stand as superlative examples of what his rooftop eagles looked like when they reigned over the skylines of Portsmouth and Kittery.

While eagles in the round, whether atop area businesses or in private gardens, were popular and profitable, Bellamy also began at this time to fill orders from other area clients for far more abstract birds. Long horizontal creations that might stretch four to twelve or more feet in length and quite akin to the sternboard eagles of the United States Navy began to appear over the entrances and across the façades of such diverse Portsmouth businesses as the H. C. Hopkins & Company Dry Goods and Furnishings Building and the grocery store of one Benjamin Norton, and music stores situated in Portland, Maine, and Manchester, New Hampshire.[7] How Bellamy went about marketing his massive abstract eagles, the warm reception his efforts received, and the generous profits he was consequently reaping are all revealed in this letter to his father, written on March 16, 1873:

> I received your letter last night. was glad to hear you are all well. I am still busy on Eagles have cut two this week for the Rev. [-?-] Hammond. They were 8 feet 6 and 12 ft 6—to be used as signs on his stores one in Manchester N.H. one to go to Portland. and have one more to make 8 ft. he has music stores in those places. I called on him one day saying that I heard of him and believed that an attractive sign would please him. Took a little one to show him so you see I made a strike in a small way. He pays me one dollar per foot—and he has got and to get including a small one 32½ feet of Eagle. I have carved one for Carne brothers to go to Boston 12 ½ ft. one for a Mr. Norton to go on his building here 12 ½ one for Mr. Prescott—and have several more under way. If I stay here a short time longer shall have them over every door as there is one now in nearly every inside. I bought yesterday enough for 7 from 12 to 14 ft—you see I make them as long as the board. so [*sic*] don't know just how long they will be. The paper men still have theirs on hand. They tell me that the heads

Fig. 6.13 Fire House at 202 Court Street, Portsmouth, New Hampshire, decorated for the 1873 "Return of the Sons of Portsmouth, NH" festivities, complete with seven-foot-long Bellamy eagle atop the arch: photograph, 1873. *UNH.*

> have *shrunk* so that they can look between the head and the wings. I tell them to set the screws up and they will be all right. Well it is now [-?-] spring and I think the eagle has done well to furnish me with comfortable quarters this winter. The Captain of the ship being built on Nobles Island is boarding here at the house—he is [-?-] the building of the ship. owns one fourth—he has offered me the work which amounts to some $300. —I told him I could not stop [to] do it. The trouble is it will cost too much—and one has to wait so long for his pay—but I have had the offer of it and I would give more for that than for the work . . . [8]

Fig. 6.13a Bellamy eagle situated on the portico of the Newfields, New Hampshire, town hall. *MD.*

Here Bellamy claims he was making so much profit so quickly that he actually turned down a lucrative contract for doing the decorative woodcarving on a merchant vessel. Having aggressively marketed them, his creations enlivened the facades of storefronts well beyond the Portsmouth-Kittery area, as far away as Boston. Bellamy's eagles were not just for storefronts, however. Local town governments also warmed to his artistic stylings, commissioning him to fashion eagles for such diverse purposes as crowning the top of an arch before a Portsmouth firehouse decorated for the "Return of the Sons of Portsmouth, N.H." festivities in 1873 [Fig. 6.13], and (possibly) [Fig. 6.13a] the portico of the town hall of the nearby village of Newfields, New Hampshire[9]. Again, this success was gained not by mere happenstance but rather by shrewd, boldly confident marketing, as when he set his sights on Portsmouth's new city hall, informing his father in a letter dated October 13, 1872: "I have made a drawing and given my figures to carve one [eagle] for the New City hall at Portsmouth for $60, it is to have a spread of 10 feet [and] think there is no doubt but what I shall get it."[10]

Unfortunately, as with so many of his large eagles in the round, time and neglect have conspired to deprive today's audience of the chance to view many of these early outdoor Bellamy eagles firsthand. A perishable art form to begin with, some were

Fig. 6.14 John Haley Bellamy, *Carved Eagle with Two Draped Flags Under a Center Shield*: painted and stained pine, ca. 1872–1890. [30 x 99 x 10 in.] *Private collection, AKA / DB.*

meant to be disposable, such as the piece crafted for the "Return of the Sons of Portsmouth, N.H." festivities, while others slowly faded and fell to the elements. And yet, like the large eagles in the round, some of these long horizontal eagles have thankfully survived.

One of these large commercial creations that remains is Bellamy's *Carved Eagle with Two Draped Flags Under a Center Shield* [Fig. 6.14]. With a length of eight and a quarter feet, this masterpiece offers us a superb idea of what the similarly sized pieces he crafted for the Rev. Hammond's music stores looked like when they first emerged from his workshop. With its sheer size and patriotic associations, such eagles would have drawn the attention of many a passing pedestrian and conferred a decisive advantage onto the business inside; its eye-catching red, white, and blue paint and gleaming gold gilding would have only increased this effect. Though the possibility that it was crafted to enliven the exterior of a private residence cannot be discounted completely, its length and the documented use of Bellamy's other similarly sized eagles strongly suggest this carving was carved for business use. Cut with flawless symmetry and possessing a lively countenance that exhibits more personality and depth of character than his usual creations, Bellamy's *Carved Eagle with Two Draped Flags Under a Center Shield* is without compare the finest surviving example of his large eagle boards.

Two striking pieces that have come to be known as the "War & Peace Eagles" also stand as perfect illustrations of the level of craftsmanship Bellamy placed into his flat, two-dimensional abstract eagles [Figs. 6.15 and 6.16]. Stretching forty-two and a half inches in length, these pine plaques bearing shields painted with the patriotic red, white, and blue color scheme may have decorated the side of someone's house or barn, though it seems more likely that they enlivened a Kittery- or Portsmouth-area storefront. Christened "War & Peace" eagles in modern times due to the persistent myth that the head of the eagle on the National Seal faces in different directions when the United States is at war from when at peace, these two birds may well

Fig. 6.15 John Haley Bellamy, *Peace Eagle:* painted pine, ca. 1872. [6 1/8 x 42½ in.] *NEA.*

Fig. 6.16 John Haley Bellamy, *War Eagle:* painted pine, ca. 1872. [6 1/8 x 42½ in.] *NEA.*

have flanked the entryway to a business. Each bird's head craning toward the front door would have served as a subtle yet effective means to capture the attention of passersby, nodding in the direction of the store's entrance and silently entreating them to walk in.

Though stylistically far removed from the three-dimensional outdoor pieces, raptors like the *Carved Eagle with Two Draped Flags Under a Center Shield* and "War & Peace Eagles" would have been equally effective in courting the attention of the fickle crowds, for though utterly lacking in realistic postures and behaviors, these plaques are successfully imbued with a sense of movement, fierceness, and regal authority. Convincing in their ability to convey the presence of the American eagle without having to rely upon realism, they stand as an effective means of delighting the eye—and as a harbinger of the next evolutionary step the eagle would take in Bellamy's art.

With the American eagle serving as his muse, Bellamy found a popular motif with lasting appeal, one that promised eager patronage and great returns. Yet despite the money he received for fashioning them, these large eagles appear to have been but a means to an end. Though their development was simultaneous, Bellamy's large birds, rather than serving as the focal point of a new venture, were instead destined to be the opening salvo of a marketing campaign designed to promote much smaller birds, wooden creatures that could be manufactured cheaply and in bulk for locals and tourists alike. Bellamy's decision to craft and sell so many oversized eagles to popular area businesses assured conspicuous display with maximum exposure of his handiwork to the general public. Increasingly,

Fig. 6.17 John Haley Bellamy, *"Don't Give Up the Ship" eagle*: painted pine, ca. 1872–1900. [9 x 25 in.] *Hyland Granby Antiques; Hyannis Port, Massachusetts. DB*. This slogan, the most common and popular used by Bellamy, was first spoken by US Navy Captain James Lawrence during the War of 1812, when his ship the USS *Chesapeake* engaged the British frigate HMS *Shannon* at the entrance to Boston Harbor.

he sought to channel the interest these pieces generated toward the sale of a product that held the promise of almost unlimited returns.

In rising to the challenge of producing a carving that was as light, compact, and easy to manufacture as it was distinctive and popular, Bellamy inadvertently conceived of a design that was as radical as it was unique [Fig. 6.17]. Bellamy's eagles had lines of a simple, fresh austerity and curves as "clean and brisk as a May morning in Kittery"[11] that conveyed an unmistakable sense of swiftness, intensity, and strength. As he had with the *Lancaster* Eagle figurehead, the *Carved Eagle with Two Draped Flags Under a Center Shield* and the "War & Peace" eagles, Bellamy achieved this not by carving with an eye toward photographic realism but by reducing details, streamlining, minimalizing, and deconstructing the American eagle down to its most basic components of head, wings, and talons. In Bellamy's hands the eagle's feathers and claws are merely alluded to, their presence achieved with a handful of bold strokes, with no attempt toward the anatomical correctness seen in his eagles in the round. Like the strains of "Yankee Doodle" frozen in mid-verse, the Bellamy Eagle hums with a brash, jaunty patriotism. The eagle "holds his proud head high, pointing a defiant tongue, a beak of great power. The eyes suggest an unquenchable ferocity; one senses an effulgence of gaiety."[12]

In crafting these birds, Bellamy was decades ahead of his time, offering us a firm hint of the expressive minimalism later found in so many of the coming century's works of modern art. Yet as forward thinking as they were, Bellamy's eagles were also thoroughly centered on the past, a product of centuries of superior Yankee ingenuity in its no-nonsense approach to making something as pragmatic in its ease of manufacture as it is beautiful to behold. While his letters show that Bellamy had

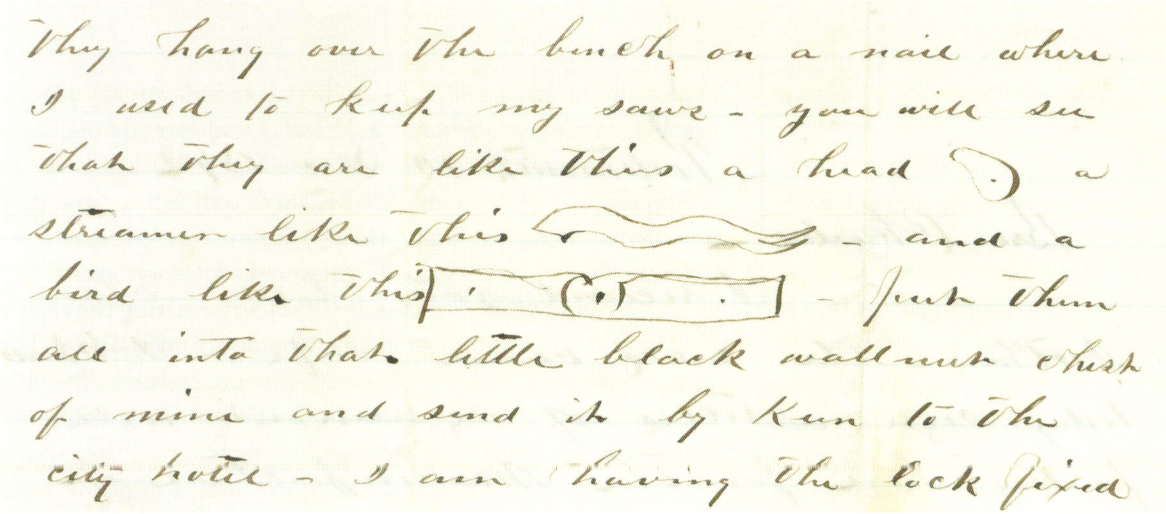

they hang over the bench on a nail where
I used to keep my saws - you will see
that they are like this a head a
streamer like this - and a
bird like this - put them
all into that little black wallnut chest
of mine and send it by Ken to the
city hotel - I am having the lock fixed

Fig. 6.18 Letter from John Haley Bellamy to Charles Carroll Bellamy, December 12, 1872, wherein John Haley Bellamy sketches out the three primary components of his famous Bellamy Eagle. *WAF*.

commenced the manufacture of these eagles in earnest by September 1873, the evolution of the Bellamy Eagle can be traced back to the artist's Charlestown shop several years before. Hints of its simple, forceful lines and the presence of a heraldic shield device can be seen within the eagle frame he produced during that time. And in the letter he had sent his brother Charles on December 12, 1872, we see that, while still living in Charlestown, Bellamy had already conceived the eagle design that would come to be synonymous with his name. In sketching the patterns Bellamy requested that his brother mail to him, he depicts three of

Fig. 6.19 A superb example of how the head assembly of a Bellamy eagle, formed out of a solid block of wood in the workshop, can be found within a series of partially carved heads in their various stages of completion, as rendered by Bellamy copyist Captain Edward H. Adams. *PPL / CO*.

Fig. 6.21 John Haley Bellamy, *"Long Live Dewey!" eagle*: painted pine, ca. 1898. [9 x 26 in.] *NEA*. This is a homage to Admiral (then Commodore) George Dewey, victor of the Battle of Manila Bay in the Spanish American War.

popular sentiments were then prevalent. Bellamy intentionally cashed in on the fierce pride and nationalism that swelled throughout America during the Spanish-American War with his "Remember the Maine!" and "Long Live Dewey!" mottos. "Don't Give Up the Ship" appears to have been targeted specifically toward tourists, as it was then emblazoned upon the famed USS *Constitution*, a popular tourist attraction in the late nineteenth century then moored at the Portsmouth Navy Yard.[27] Similarly, his "*Nil desperandum Christo duce*" ("Despair of nothing while Christ leads") eagles display a Latin motto that was the battle cry of Sir William Pepperrell's colonial forces in their capture of the French fortress of Louisburg in 1745. They were made for those attending festivities on the one hundred and fiftieth anniversary of Pepperrell's successful campaign in 1895. Indeed, it appears that those eagles adorned with this saying may have been crafted specifically with an eye toward tourists that were visiting the Pepperrell Mansion at that time.

As easily as his raptors offered a proud cheer to their nation's health and heroes, so could they be employed toward wishing people a merry Christmas [Fig. 6.22], a happy Thanksgiving,[28] or a happy New Year [Fig. 6.23]. While not as common, purely religious messages appeared from time to time, quoting popular scripture passages such as "The Lord is My Shepherd" [Fig. 6.24]. Undoubtedly crafted on commission, some Bellamy eagles could be found bearing the initials of various fraternal organizations in the Portsmouth/Kittery area, like the "FPVFA" lettering in the banner atop the portico of the Franklin Pierce Veterans Firemen's Association building, [Fig. 6.25] and the Wentworth K. P. inscription for an eagle once owned by the Wentworth Lodge of the Knights of Pythias in New Castle, New Hampshire [Fig. 6.26]. Banners in Latin appeared with varying frequency, not only sacred maxims like the "Nil desperandum Christo duce" eagles, but purely secular sentiments like "*Dum Vivimus Vivamus*" ("While we live, let us live") [Fig. 6.27] the famous, all-too-worldly and materialistic assertion of Harvard College's exclusive Porcellian Club. Some eagles possess banners whose inscriptions make little sense today, like the mysterious "RENIM," [Fig. 6.28] which could possibly be a Latin

Fig. 6.22 John Haley Bellamy, *"Merry Christmas!" eagle*: painted pine, ca. 1872–1900. [9 x 25 in.] *HG / DB.*

Fig. 6.23 John Haley Bellamy, *"Happy New Year!" eagle*: painted pine, ca. 1872–1900. [9 x 25 in.] *KHNS. Photo by Donald Chick.*

Fig, 6.24 John Haley Bellamy, *"The Lord is my Shepherd" eagle*: painted pine, ca. 1872–1900. [9 x 25 in.] *Hyland Granby Antiques; Hyannis Port, Massachusetts. DB.*

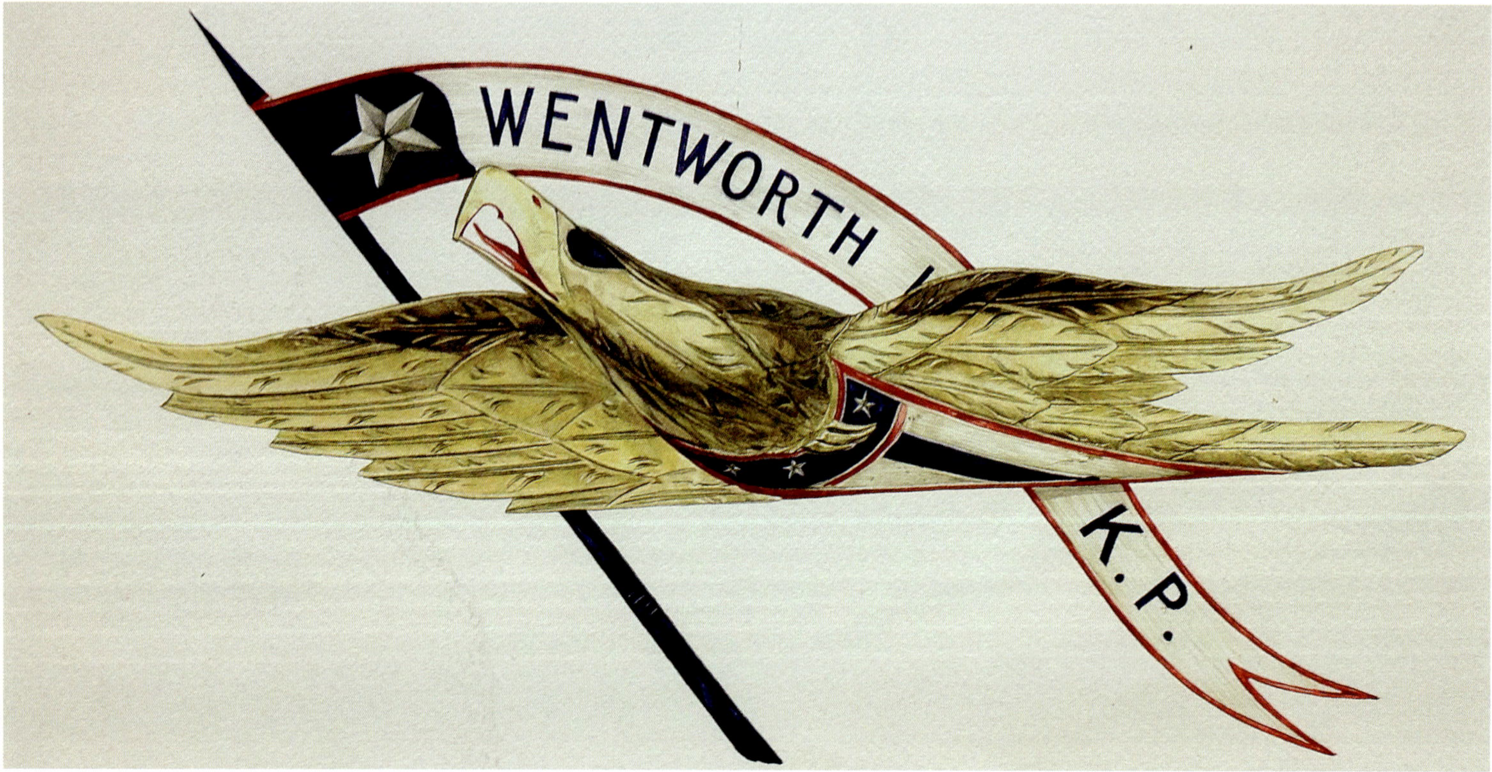

Fig. 6.26 Alice Cosgrove, *Rendering of a Carved Eagle*: watercolor and graphite on paper, 1940. [11 x 20 9/16 in.] *NGA*. Painted by artist Alice Cosgrove for the Index of American Design in 1940, the whereabouts of the original carving is now unknown.

Fig. 6.25 Bellamy eagle with letters *F.P.V.F.A.* on the eagle's banner, located above the door on the Franklin Pierce Veterans Firemen's Association, Portsmouth, New Hampshire [inset]: photograph, ca. 1905. *Private collection*. Note the letters *F.P.V.F.A.* inscribed within the eagle's banner.

Fig. 6.27 John Haley Bellamy, *"Dum Vivimus Vivamus" eagle*: painted pine, ca. 1872–1907. [9 x 24½ in.] *Private collection. CO.*

Fig. 6.28 John Haley Bellamy, *RENIM. Eagle*: painted and gilded pine, ca. 1872–1900. [9 x 24½ x 3¼ in.] *Ronald Bourgeault. CO.*

Fig. 6.29 John Haley Bellamy, *Mahogany eagle*: mahogany, ca. 1872–1900. [7¼ x 34 in.] *KHNS / DB.*

abbreviation. A small number were also made of mahogany, though today such examples are exceedingly exquisite and rare, [Fig. 6.29] and offer no clear reason for their having been carved out of this more expensive wood.

Bellamy's two-foot-long eagles were christened with so many different sayings, expressions, and mottos because of the differing tastes of his clientele. By appealing to a host of patriotic, religious, and fraternal sentiments, Bellamy's handiwork was assured a wide and diverse audience. Just how wide this audience was Moses Victor Safford was at a loss, admitting, "Where he found a market for these smartly decorative carvings it is hard to say. I have seen him boxing such wares for shipment; but so many small eagles were presented to friends that their making seems to have been primarily a diversion."[29] However, it is now plain that making eagles was anything but a diversion for the carver of Kittery Point. As various testimonies and newspaper clippings demonstrate, producing eagles was pure business for Bellamy, one that paid him handsomely. As the Kittery newspaper editor and publisher Horace Mitchell once wrote, "Orders for eagles were numerous; one was for 700. Bellamy's usual price was about one dollar each."[30] Mitchell's claim that Bellamy supplied seven hundred eagles is supported by other sources that reveal that it was common enough for the woodcarver to fill such large requests. As described in the *Portsmouth Daily Chronicle* on September 2, 1885:

> "John Bellamy, of Kittery Point, at his workshop in the rear of the old Pepperrell mansion is busily engaged in carving wooden eagles. He has recently executed 500 effigies of this proud bird for Frank Jones."[31]

It has been noted that the eagles from his small workshop "must have been widely distributed." Indeed, John Haley Bellamy and his assistants also supplied his famous birds to markets and clients well beyond the Piscataqua. The occasional Bellamy eagle may still be found with a sticker fixed on its back naming The Manhattan Storage & Warehouse Company, confirming that Bellamy shipped his eagles and marketed them to a New York-based clientele and beyond. Such shipments were what Safford beheld when he witnessed his uncle "boxing them and shipping them away by express from his Kittery Point shop" (MVS: *Eagle*, Pg. 9). They also reveal that Bellamy's reputation stretched well beyond Kittery. Safford claimed that this rested solely upon his uncle's hard-won standing as an exceptional carver. "His reputation brought him orders from other sources. He seems to have made little or no effort to solicit commissions."[32] So successful was Bellamy in securing orders that he once admitted "I could always find the market for my doings."[33]

Reputation alone, however did not sell his creations. Bellamy's predilection for roundly distributing eagles as gifts to friends and neighbors was probably more than just a gesture of

Fig. 6.30 Frank Jones, entrepreneurial and political dynamo of Portsmouth, New Hampshire, and the most lucrative of John Haley Bellamy's business partners: oil on canvas painting by Nellie Mathes Horn, ca. 1880. [30 x 25 in.] *PA*.

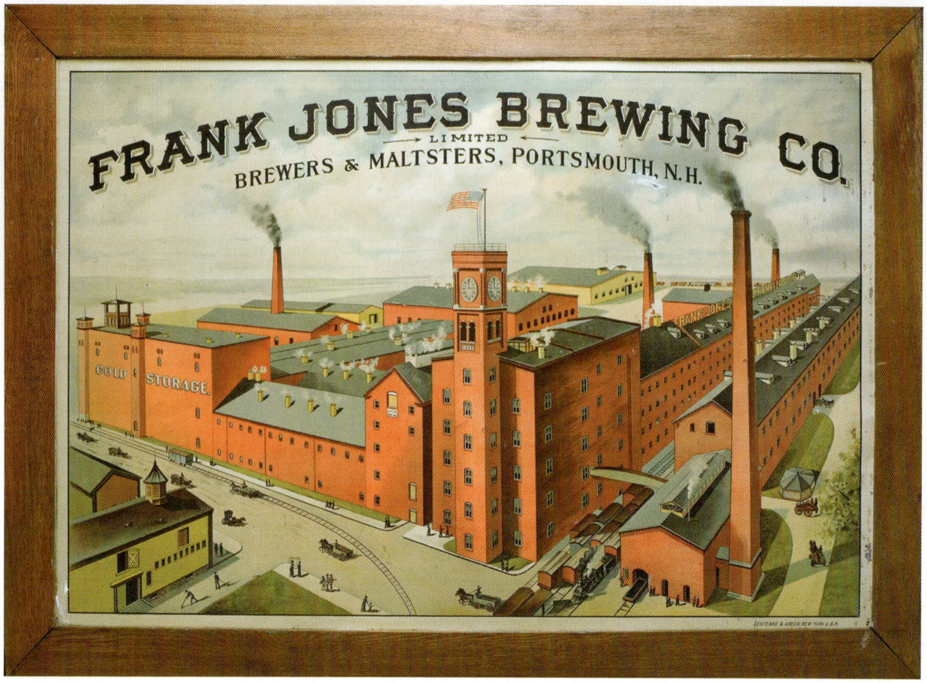

Fig. 6.31 "Frank Jones Brewing Co. Ltd. / Brewers & Malters / Portsmouth, NH": advertising lithograph on tin, ca. 1890. *PMS / RM*. This business venture was the bedrock upon which Jones established his financial empire and the probable source for the distribution of many of Bellamy's holiday-themed eagles.

openhanded kindness; it was an effective tool to keep his creations visible in the public eye and build up favorable word of mouth about them. He made certain highly visible local institutions and businesses sold his distinctive eagles, such as the substantial number of his distinctive eagles he produced for Frank Jones.

Railroad baron, hotelier, insurance magnate, steamship operator, banker, owner of public utilities, senator, personal friend of United States presidents, and owner of one of the largest and most successful ale breweries in the United States, Frank Jones was a dynamo in the business affairs of Victorian-era Portsmouth and Kittery [Figs. 6.30 and 6.31]. From a December 12, 1872 letter to his brother Charles, we know that he was already doing business with Portsmouth's premier captain of industry, for Bellamy makes mention of "the patterns of the birds father and I made for Jones." Bellamy was then boarding at Portsmouth's City Hotel, an institution owned by Jones,[34] and given that his "partners here for the manufacture of eagles" were specifically paying his board at the City Hotel, one can even speculate that Frank Jones was among those initial investors.

By doing a brisk business with Portsmouth's premier commercial tycoon Bellamy was assured generous patronage and a large number of venues for broadcasting his talents to the general public. The most plausible use Jones found for such a large

Fig. 6.32 Interior view of an unidentified Portsmouth bar displaying a Bellamy eagle plaque. Furthering the assumption that Bellamy's eagles and eagle plaques were intimately tied into the promotional ventures of the Frank Jones Brewing Co. Ltd. is the presence of a sign hanging above and to the right of the unidentified bartender that reads "We sell Jones & Eldredges Portsmouth Ales. State Your Choice." *PA*.

numbers of Bellamy eagles appears in the sayings emblazoned upon them. Holiday greetings of "Merry Christmas" and "Happy New Year" are rather common when reviewing the long list of sayings Bellamy placed on the banners and Jones appears to have liberally distributed his eagles near the holidays as a promotional venture. This would not only account for the large number of the eagles sporting holiday greetings but why so many of Bellamy's eagles found their way into local Portsmouth taprooms and taverns, given that Jones' brewery was their primary supplier [Figs. 6.32].

To fill orders of the size he was taking on for the likes of Jones and others, capable assistants and an easily assembled product were two necessary factors for Bellamy to meet his production deadlines. Also essential for success was an efficient, practical workspace. Located right in his back yard, Bellamy's "snug little factory" was situated in the former counting house of Sir William Pepperrell, an ancient dockside office that afforded the artist ample space and light, with sweeping views of the wharves and harbor [Fig. 6.33]. Furthermore, the old counting house was ideally situated for acquiring and storing stock purchased on the nearby wharves and for shipping crates of finished carvings to distant markets. Used in years past by his father as a boatbuilding shop, Bellamy probably set about converting the second floor into his workspace upon his return from Charlestown in 1872. A priceless photograph of the studio taken at the height of its activity shows it was a cluttered affair, filled with tools and various key examples of Bellamy's handiwork (leading one to assume that the photo was staged and taken for promotional purposes) [Fig. 6.34].

For a deeper appreciation of what it was like to stand inside Bellamy's workspace, we may rely on Moses Victor Safford's description:

> [The] shop was a single room in the second story of the building where Bellamy's father had kept his contractor's gear and, with his other sons, had built boats. Its windows opened toward the harbor and the sea to the south. From them at times the eye might catch the looming Cape Ann in the distance. A ship's ladder with a manrope led up to the shop. An iron weight on the end of a rope, which passed through an old clew-line block, closed the door behind the visitor. Along the south side of the shop extended a workbench. Above it hung recent carvings and many patterns and tools used in the woodcarver's craft. The walls were, for the most part, papered with yellowing drawings by Thomas Nast and with brighter-colored political cartoons from *Puck*. Strung about here and there was a varied assortment of prints, carvings, drawings, and curios. A pot-bellied stove standing in a box of sea sand supplied genial warmth during the colder months of the year. Comfortable chairs of diverse styles and ages were placed here and there to accommodate visitors.[35]

Elsewhere, Safford elaborated on this description, writing that the many curios in the shop were drawn ". . . from every where [*sic*] in the world" (MVS: *Eagle*, Pg. 8). One item Safford neglected to mention was Bellamy's stuffed taxidermied eagle,[36] an item that was probably not so much a curio as it was a research tool the artist turned to when engaged in crafting the occasional realistic eagle in the round, like the piece perched on the far left of Bellamy's workbench in the studio photograph.

OF BELLAMY HIMSELF THERE IS NOTHING MORE THAN a few photographs and stories from which to reconstruct his disposition when in the workshop. A photograph of the artist from the beginning of this period in his life reveals him to have been a hearty, hale Maine Yankee, a seasoned man gazing down at the camera with folded arms and beady, uncompromising stare [Fig. 6.35]. As to the nature of the man behind those dark, piercing eyes, it has been claimed that, when involved in his work, "Nobody ever saw Bellamy sit down in his workshop, except to rest for a few minutes on the high stool that he kept near him at the bench."[37] He "was an industrious and rapid worker. Visitors

Fig. 6.33 View of John Haley Bellamy's workshop from Sir William Pepperrell/West Indies Wharf, late ninteenth/early twentieth century. Bellamy's workshop is the small two-story structure with open second floor windows dead-center in the photo. The Sir William Pepperrell Mansion stands immediately behind it and to the left. *Private collection.*

were courteously received but they never caused him to stop work entirely and lay aside anything that he was doing" (MVS: *JHB*, Pg. 12).

Industrious and naturally driven, Bellamy had the right balance of artistic skill, business savvy, focus, and forceful personality to excel within his chosen profession. This heady mixture of traits would also ensure that he did not grow complacent, churning out two-foot-long banner-toting flat eagles for the remainder of his days. Whether born out of the demands of customers, the determination to broaden and expand his profit margins, or the nameless desire to innovate and invent, John Haley Bellamy began to produce and fill orders for elaborate, distinctive eagle plaques, while still crafting and distributing his trademark birds by the hundreds. Unlike with his classic

Fig. 6.34 John Haley Bellamy's workshop, featuring his workbench and four various examples of his carvings: photograph, ca. 1872–1900. *WAF.*

Fig. 6.35 John Haley Bellamy: tintype, circa 1876. *WAF.*

two-foot-long raptors, there is as yet no documentation to his having produced his eagle plaques in heavy volume or distributed them to international markets. And though they bear the telltale marks of scroll saws that betray themselves to have been produced quickly and in number, a greater divergence in design and construction can be found within these creations. This may suggest they may have been less than a staple product of Bellamy's workshop. Nevertheless, the eagle plaques were a substantial facet of his business, as attested to by the many surviving examples known to still exist.

The basic characteristics found in a classic Bellamy eagle plaque [Fig. 6.36] involve a relatively flat wing assembly akin to, yet distinctly different from, a typical Bellamy eagle, with the wings spread wide in a less dynamic, more formal, almost ceremonial posture. The presence of a richly painted shield and brace of crisscrossed American flags are almost always found, and a motto-bedecked banner is often clenched fast in the eagle's beak, though less elaborate examples of this type of eagle do exist [Fig. 6.37]. Invariably when viewing Bellamy's eagle plaques, one is reminded of heraldic devices, such as the Great Seal of

Fig. 6.36 John Haley Bellamy, *"God Is Our Refuge and Strength" eagle plaque*: painted and gilded pine, ca. 1872–1900. [17 x 48 in.] *Private collection. CO.*

Fig. 6.37 John Haley Bellamy, *Eagle plaque*: painted pine, ca. 1872–1900. [9 x 24½ in] PMS, RM. Note the absence of a banner, the smaller, perfectly symmetrical shield, and the overall subdued, quiet presence of this eagle plaque when compared to other, more ebullient Bellamy eagle plaques.*Private collection. CO.*

Ticonderoga

No. E-31 31″ x 18″

$140.00

All dimensions give the overall width or wingspread first.

John Quincy No. E-35 35″ x 6½″ $35.00

Streamer No. E-45 45″ x 7″ $34.00

Mt. Vernon

No. E-19.1 19″ x 9″ $33.00

Primitive

No. E-19.2 19″ x 10″ $50.00

Portsmouth No. E-28 28″ x 9″ $35.0

Salem

No. E-25

25″ x 9″

$50.00

The handsome eagles shown on these three pages are accurate reproductions of authentic Early American originals, done in the tradition of fine craftsmanship. All carving is hand finished in seasoned pattern pine. The prices shown are for eagles brilliantly decorated with compo gold leaf and authentic colors*, to be as inspiring to you as the originals were to our forefathers.

*Eagles are available in karat gold leaf at 20% additional, and in a variety of woods and custom finishes on special order. Send specifications and Style No. for quotation.

2

Fig. 7.9 Examples of Artistic Carving Company of Boston copies of Bellamy eagles. *NEA.*

end, when the eagles were assembled and gilded, painted, and antiqued. Auctioneer Ronald Bourgeault remembers his early introduction to the company's Boston factory.

> When I was 10 years old and worked for Webber's Antique Shop in Hampton, we would get in the green pick-up truck and drive down to Boston, next to North Station, where there was a company called Artistic Carving Company. As a kid I saw a multiple lathe about the size of that wall, worked by one man carving about a dozen eagles at a time. He'd do the backs, then the head, and then they'd be assembled. A few hand-carved strokes were put on and the eagles would go to the "antiquing" room, where they'd get a dark varnish. That was my first exposure to what were purported to be "Bellamy" eagles. Occasionally an old one would show up at Webber's, but very rarely and usually in poor condition.
>
> . . . . Over the years Artistic Carving Company eagles have "opened their wings" and made it into many, many different auction catalogs, sometimes bringing phenomenal amounts of money because people do not know about them. I finally found one of their old catalogs and reproduced a few pages in my 2005 Marine Auction catalog.[7]

Fig. 7.8 (left) Advertisement for the Artistic Carving Company of Boston, ca. 1950. *NEA.*

Sporting such names as the "Portsmouth," the "Streamer," the "Mt. Vernon," and the "John Quincy," the pieces directly sought to copy both Bellamy's conventional two-foot-long eagles [Fig. 7.9] and his more elaborate eagle plaques.

Fig. 7.16 (top left) Joseph True, *U.S. Customs House Eagle (interior), Salem, Massachusetts*: gilded wood, ca. 1826. [12 x 39 x 16 in.] *NEA*.

Fig. 7.17 (top right) Laban Beecher, *Architectural American Eagle*: gilded wood, n.d. [56 x 39½ in.] *NEA*.

Fig. 7.18 (left) George Stapf, *Carved and painted eagle plaque with shield and American flag banners*: painted wood, n.d. NEA. For many years numerous authorities in American antiques and fine art have insisted upon misattributing George Stapf's handiwork to John Haley Bellamy. (The most famous instance of this was lot 875 of Sotheby's Garbisch Sale, on May 22–25, 1980.) As this typical Stapf eagle clearly illustrates, there are no similarities whatsoever to be found between the two carvers'works.

Fig. 7.19 Photograph of Kittery Point resident Windam Mayo, circa 1900, with a painted Ivah W. Spinney wall eagle mounted over his doorway. Aside from the fact that their heads are turned in opposite directions and one is painted while the other is gilded, the similarities between this Spinney eagle and the *Carved gilded wall eagle* (Image 139) are striking, hinting that Spinney possibly made use of similar production methods as Bellamy. *Private collection*

However, some local area artists managed to produce in sufficient number several competent eagle carvings quite akin to Bellamy's style.

One of these inspired Bellamy copyists was Ivah W. Spinney, a native of nearby Portsmouth, New Hampshire, who was actively carving at the same time as Bellamy. Spinney's carvings demonstrate an obvious (if not shameless) debt of influence to Bellamy's eagles. Because Spinney was carving during the same time as his famous competitor across the river in Kittery, both artists' eagles are roughly the same age and both thus sport period hardware and heavy oxidation upon their backsides. And as with so many Bellamy eagles, Spinney's were to be found gracing many a home in the Portsmouth and Kittery area [Fig. 7.19]. Differences do exist, however.

While Spinney's birds are indeed highly stylized, they often take on forms unknown among Bellamy's artistic oeuvre [Fig. 7.20]. The shields found on Spinney's large wall plaques can look more like a Grand Army of the Republic armorial device than anything a Bellamy eagle has ever held fast in its talons [Fig. 7.21]. The feathers carved into a Spinney eagle are generally of a softer, shallower cut and hold no immediate similarities with Bellamy's handiwork; the "lift" he tries to replicate along the top edge of the wings is too brief and shallow when compared to Bellamy's eagle wings. And while Spinney did engage in the occasional copy of a typical two-foot-long Bellamy eagle, [Fig. 7.22] complete with highly accurate copying of Bellamy's signature treatment of the feathers and shield, they are distinguishable from the original product by the fact that their necks lack an authentic Bellamy eagle's telltale graceful upward sweep; the necks of Spinney's birds often run fairly parallel with the top of the wing assembly. The wings themselves differ considerably from Bellamy's, given their angular, "awkward structure," while "the laboriously hard lines of gouging, and the absence of the crisp, light-catching accents with which Bellamy enlivened his work"[11] also serve to separate Spinney's copy work from Bellamy's original compositions. Lastly, a Spinney eagle may even feature the stamp of its creator running along the top of

Fig. 7.20 Ivah W. Spinney, *Carved Gilded Wall Eagle*: painted and gilded wood, early twentieth century. [12 x 40 x 6 in] *Private collection. CO.*

Fig. 7.21 Ivah W. Spinney, *Large Hand-Carved American Wall Eagle*: painted and gilded wood, late nineteenth century. [75 x 16 x 6½ in] *Private collection. CO.*

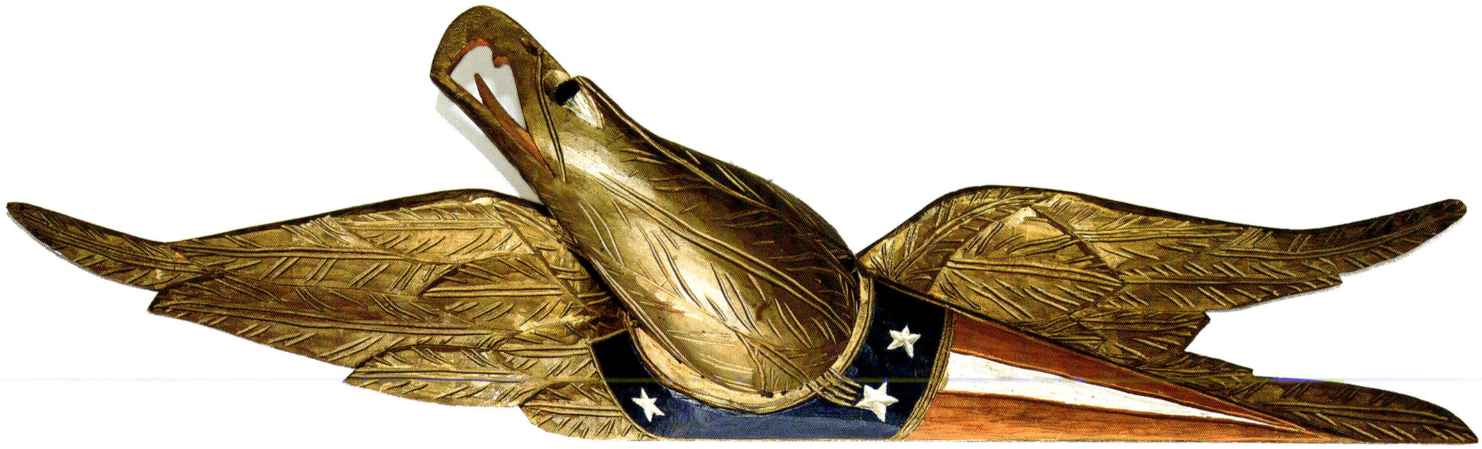

Fig. 7.22 Ivah W. Spinney, *Copy of a Bellamy Eagle*: painted and gilded wood, late nineteenth century. [9 x 25½ in] *Private collection*.

Fig. 7.23 Detail of Spinney's signature appears atop the edge of the bird's left wing. *CO*.

the wing, banishing all doubt as to who crafted it into being [Fig. 7.23].

Two other prolific copyists contemporary to Bellamy were a Captain Edward H. Adams and his son Cass. Residents of Adams Point in nearby Durham, New Hampshire, this father and son team, having built and sailed the last gundalow ever to grace the Piscataqua River, whittled a wide assortment of carvings, ranging from birds, fish, and ship models to remarkably adept copies of the classic Bellamy eagle. Derived from an in-depth study of Bellamy's handiwork (Captain Adams claimed to have purchased a Bellamy eagle from John Haley Bellamy personally), Edward Adams hand carved his eagles; his son Cass applied the paint that brought them to life. Thoughtfully executed, the final product can often be discerned from an actual Bellamy eagle by taking notice of the subtle, bulbous imperfections in the lower portion of the head/neck assembly, the atypical, less-accomplished lettering found within the banner, and the negligible yet

Fig. 7.33 Plaster cast of John Haley Bellamy's *Maine State Seal*. Original ca. 1872–1900. [15 in. diameter] *MM.* Cast from a sand mold carved out of wood by Bellamy, this seems to be a copy of one of a pair reputed to have hung first in Sparhawk Hall and then in the Bellamy /Pepperrell Mansion.

as facile and adept at replicating modern ceremonial devices, including the seal of the State of Maine [Fig. 7.33].

An often-overlooked aspect of Bellamy's career is his labor as a decorative house carpenter. Known to have engaged in this work upon his return to the Kittery area in the early 1870s, Bellamy unquestionably learned this particular function of the woodcarver's trade from his father, "an architect of excellent ability and good taste."[20] Today many residences in the Kittery/Portsmouth area still sport decorative interior accents carved by Bellamy himself; more than a few homeowners are not even aware that they are so fortunate as to have his workmanship gracing their house. Bellamy's handiwork might appear nearly anywhere in the fabric of a home, inhabiting dark corners and high-traffic areas alike with a quiet, unassuming dignity. What usually first serves to alert homeowners that they have a Bellamy carving is the way a particular piece of woodwork will gently stand out from its surroundings, such as the acanthus leaf newel post on an otherwise plain and unadorned staircase railing found in a Kittery home once inhabited by Bellamy's niece, Mary Safford Wiles. Its unique design and delightfully surprising attention to detail, [Figs. 7.34–7.36] makes it evident that the newel post was made by hand by an expert carver. From its distinctive, languid *S*-shaped cuts and use of stippling to add an illusion of depth and texture—two traits characteristic of many a Bellamy sculpture—the identity of its creator is firmly established. Upon inspection, the piece, crafted out of American chestnut, reveals that it was not carved *in situ* but was crafted in Bellamy's shop and joined to the otherwise generic banister upon completion.

Another decorative accent Bellamy fashioned within his shop prior to installation were his pine laurel wreaths [Fig. 7.37]. As with his acanthus leaf newel post, these carvings remind us that the great woodcarver's creations could appear almost anywhere, hiding in plain sight, as they did for decades untold on the portico frieze of Portsmouth's Governor Ichabod Goodwin Mansion [Fig. 7.38]. Yet of all Bellamy's house carving commissions, perhaps no residence benefited as much from his hands as Kittery's beloved and much lamented Sparhawk Hall [Fig. 7.39].

Originally constructed in 1742 as a dowry gift from Sir William Pepperrell to his daughter Elizabeth and her new husband, Nathaniel Sparhawk, this picturesque mansion "was fortuitously scooped up" in the year 1866 by Mr. Walter Brown, a successful banker from Toronto, Canada, who sought throughout the last half of the 1800s to diligently preserve the house in what may well have been "the first colonial private home in America to be consciously 'restored.'"[21] Brown, whether out of a personal liking for Bellamy's creations or merely from trusting in his reputation as a superb artisan, hired both the great woodcarver and his father around the year 1868, at a time when Bellamy was still spending the majority of his days in Charlestown, for the task of renovating his new home. Perhaps because of their presumed

Fig. 7.34 (top left) Detail of John Haley Bellamy's *Acanthus leaf newel post*: American chestnut, ca. 1870–80. [10 x 39 in.] *Private collection. KC, 2008.* Of particular note are the similarities between the *S*-shaped marks running along the top of the post and those found running along the top edge of the wings on many of Bellamy's eagles.

Fig. 7.35 (bottom left) Detail of John Haley Bellamy's *Acanthus leaf newel post*: American chestnut, ca. 1870–80. [10 x 39 in.] *Private collection. KC, 2008.* The intense devotion with which Bellamy textured the wood on this post makes this a particularly masterful creation, one all the more extraordinary given its mundane, functional nature.

Fig. 7.36 (above) Detail of John Haley Bellamy's *Acanthus leaf newel post*: American chestnut, ca. 1870–80. [10 x 39 in.] *Private collection. KC, 2008.*

Fig. 7.37 John Haley Bellamy, *Pine laurel wreath*: pine, ca. 1875. [12 x 12½ in.] *WAF / DB.*

7.38 Pine laurel wreaths by John Haley Bellamy adorning the entrance portico of the Governor Ichabod Goodwin Mansion, Strawberry Banke Museum, Portsmouth, New Hampshire. *CO, 2013.*

knowledge of colonial woodwork gained from their life in the Pepperrell Mansion, Brown appears to have given the two men great latitude. The result was a Bellamy *tour de force*, a magnum opus where father and son used Sparhawk Hall [Fig. 39] as their own personal canvas and in the process filled the house to the point of bursting with their handiwork.

Interior photos shot soon after their work on Sparhawk Hall was completed show the variety of workmanship Bellamy and his father installed throughout the property. In the front hallway [Fig. 7.40] we encounter three distinct pieces of John Haley Bellamy's handiwork. At the head of the stairway, hiding in plain sight in a fashion no different from his other interior decorative treatments, is a wooden pineapple carved in the round by the artist, fixed squarely atop the banister post. Suspended from the ceiling in the upper-center portion of the photograph a unique Bellamy creation can be discerned, a carved hawk, while in the extreme upper right corner of the photograph we catch a glimpse of a moose head also cut by him.

Fig. 7.39 Famed Sparhawk Hall, as it appeared in the years immediately prior to its destruction. *PA*.

Fig. 7.40 The front entryway of Sparhawk Hall, complete with multiple Bellamy sculptures. *PPL*.

After Sparhawk Hall was deconstructed and burned "like an old worn flag" in 1967,[22] only one of these three articles is presently known to have survived. The pineapple has fallen into obscurity, possibly destroyed when the house was set aflame, while the moose head, which we know to have been removed from the hallway prior to the home's destruction, has since disappeared, leaving today's audience with only one precious up-close photo of the piece from which to gain an appreciation of its size and beauty [Fig. 7.41]. Fashioned out of black walnut,[23] the moose head was identical in purpose to the previously discussed deer heads, in that it served as a hunting trophy; Bellamy incorporated the actual antlers of the moose that was killed into the carving. (Such a sculpture may also bring up-to-date a local tradition of displaying horns and antlers in the eighteenth-century hallways of the local elite.)

The only known survivor from the main hallway is the hawk. Now safely within the collection of the Portsmouth Athenaeum, [Fig. 7.42] this piece is another of those "unfamiliar

Fig. 7.41 John Haley Bellamy, Moose head: black walnut, ca. 1868. [measurements unknown] *Courtesy of Scholl Antiques.*

Fig. 7.42 John Haley Bellamy, *Hawk-on-a-Spar*: black walnut, ca. 1868. [22½ x 13 x 11 in.] *PA / CO.*

7.43 John Haley Bellamy, side view of Hawk-on-a-Spar: black walnut, ca. 1868 [22½ x 13 x 11 in.] *PA / CO.* As with his *large American eagle atop an out-of-round ball*, we find the presence of a dramatic profile and puffed-out chest.

productions," in fact the only known depiction of a bird other than an eagle that Bellamy ever made. Cut from black walnut and composed of five pieces, Bellamy's "Hawk-on-a-Spar" is an obvious play on the name of the house in which it hung, leaving us to wonder if it is a product of Bellamy's imaginative wit or his patron's. Remarkably akin to his *large American eagle atop an out-of-round ball*, particularly the sweep of its tail, its profile, and its delightfully puffed out chest, [Fig. 7.43] it is a charming sculpture, filled with the "overwhelming *joie de vivre*"[24] found throughout so much of his work.

Fig. 7.44 The surviving paneling and shell cupboards of Sparhawk Hall's parlor as they appear today. (The cupboard on the left is the handiwork of John Haley Bellamy, while the shell cupboard on the right is original to the house, crafted ca. 1742. *D & C /CO.*

Within Sparhawk Hall's parlor a completely different side of Bellamy's carving skills is revealed, along with an indication of just how closely he and his father collaborated on their woodworking projects. The paneling that once adorned the grand home's parlor was *not* engulfed in the 1967 destruction of this famed abode but was fortuitously removed in 1953 and later installed by the Woolworth family inside their Augusta, Maine, estate. [Figs. 7.44–7.45].[25] Behind the panels directly above the fireplace is the signature of his father [Fig. 7.46] with the inscription, THIS PANEL REMOVED BY ROBERT SAFFORD. REPLACED BY CHARLES G. BELLAMY, JANUARY 27, 1868.[26]

The parlor of Sparhawk Hall stands as a priceless testimony to the intense, seamless level of collaboration that existed between Bellamy and his father as they labored together on the woodwork, adorning the same room in the same house at the same time. We are also afforded the chance to realize that John Haley Bellamy was capable of moving beyond the conventional carvings relative to his time, producing first-rate reproductions of decorative woodwork from centuries-old traditions that were virtually indistinguishable from the original, attesting to how versatile and skilled he was when he stepped beyond the parameters of his usual handiwork. This is especially true relative to the room's left-hand shell cupboard: "Well-known ship-carver John Haley Bellamy, of Kittery, was hired to replicate a shell-headed

Fig. 7.48 (top) Schooner *Mary Willey*, with Captain Williams and crew, photograph, August 27, 1896. John Haley Bellamy is the last person on the far right. *WAF*.

Fig. 7.49 (middle) John Haley Bellamy, *Quarterboard for the schooner* Mary Willey*:* painted and gilded pine, ca. 1896. [6½ x 73½ in.] *KHNS / DB*.

Fig. 7.50 (bottom) John Haley Bellamy, *Quarterboard for the schooner* Golden Gate: painted and gilded pine, ca. 1869–1900. [7½ x 108 in.] *KHNS / DB*.

Fig. 7.51 John Haley Bellamy, *"Don't Give Up the Ship" eagle for the schooner* Gracie D. Buchanan: painted pine, ca. 1869 [5¾ x 16 in.] *Olde Hope Antiques, Inc.*

Fig. 7.52 The tugboat *Mitchell Davis*: photograph, n.d. *WAF*. Note the eagle by John Haley Bellamy perched atop the vessel's pilot house.

Ever adaptable, Bellamy also crafted works for contemporary vessels, such as the masterful full eagle in the round he carved for the steam-driven tugboat *Mitchell Davis*, commanded by Bellamy's friend and neighbor Captain Hoyt [Fig. 7.52]. Still in existence today, the pilothouse eagle is notable in how it demonstrates a genuine sense of lift within its wings, conveying a dynamic, lively energy with a minimum of detail, much akin to the *Lancaster* figurehead [Fig. 7.53]. Sensitively rendered from the tip of its wings to the very last tail feather, this striking example has been declared "the finest example of his three-dimensional carving,"[30] and confirms yet again how accomplished Bellamy was in crafting a true-to-life three-dimensional carved image of the American eagle.

Remembering that Bellamy's training began with Portsmouth's Samuel Dockum, a man who was among other things an accomplished furniture maker, we might expect to encounter a good number of Bellamy's furniture pieces. However, examples of this facet of the great woodcarver's career remain surprisingly rare, no doubt because many pieces were disposed of over the years when changes in fashion called for their replacement by more modern furnishings. Among the scant pieces that have been unearthed over the years, including a walnut settee and a pine shelf, one particularly unusual and beautiful carved walnut folding shelf by Bellamy has surfaced. The shelf, with its diamond-shaped back and splendidly convincing carved ropes and tassels, [Fig. 7.54] is not only a minor Bellamy masterwork but is completely devoid of any Masonic or Odd Fellows symbols, informing us that Bellamy had crafted what-not shelves for the private market as well.

Equally rare are those stray bits of whittling that can be regarded as products of Bellamy's eccentric ways. Whether cut out of boredom, as experiments, for utilitarian purposes, or for sheer amusement, in the true tradition of a Maine woodworker Bellamy fashioned a number of unique items over the years intended for his and his family's use. A one-of-a-kind decorative

Fig. 7.53 John Haley Bellamy, *Pilothouse eagle for the tugboat* Mitchell Davis*:* gilded pine, ca. 1890. [32 x 40 in.] *Private collection. DB.*

Fig. 7.54 John Haley Bellamy, *What-Not shelf*: walnut, ca. 1868–1900. [8 x 13 x 7½ in.] *NEA.*

plaque containing a motif of roses carved by Bellamy around 1864 [Fig. 7.55] was once displayed outside the Pepperrell Mansion; it appears to have been crafted purely for the enjoyment of the Bellamy family and their neighbors [Fig. 7.56]. Meanwhile, a mahogany pen fashioned in the shape of a quill and a set of miniature books cut from teak and mahogany taken from the Spanish warship *Reina Mercedes* (a cruiser that was captured during the Spanish-American War by the American Navy and repaired at the Portsmouth Navy Yard) appear to have been little more than flights of fancy [Fig. 7.57].

Perhaps the most striking example of this type of eccentric carving that Bellamy ever produced involves a certain mahogany box [Fig. 7.58]. From an inscription on the bottom of the piece in the artist's own hand, we discover that he crafted this box in 1876, the American Centennial, when patriotic fervor was running high and the public clamored for anything bearing the trappings of national authority [Fig. 7.59]. Festooned with carvings, the richly ornate piece bears carvings upon its panels that resound with whimsy of both a nautical and pure Americana focus. Upon the top of the box a crisp, superb carving of the Great Seal of the United States sits amid a swirl of flawlessly executed botanical vines and leaves [Fig. 7.60]. Upon the front panel a banner-and-shield-toting eagle, akin to Bellamy's wall plaques, is fixed within a similar tangle of vines, [Fig. 7.61] as is a solitary shield upon the back. [Fig. 7.62] The side panels feature two identical, quasi-angelic figures that would appear to have been inspired by the same maritime tradition that guided Bellamy in his childhood painting *The Flying Dragon* [Fig. 1.6]. Even the feet of the box reverberate with Bellamy's enchanting eccentricity; he having rendered them into tiny paws and thus anthropomorphized the wooden chest into a living creature [Fig. 7.63]. With its distinctive design and the sheer amount of detail,

Fig. 7.55 (far left) John Haley Bellamy, *Carved plaque with motif of roses*: painted pine, ca. 1864. [10½ dia.] *WAF/ DB.*

Fig. 7.56 (left) A close-up view of the side door to the Sir William Pepperrell Mansion, taken in the late 1800s, revealing the carved plaque with motif of roses (then coated in a dark paint or stain) positioned dead center in the portico, with the Laban Beecher eagle hovering right below it. *DK.*

Fig. 7.57 (bottom) John Haley Bellamy, *Wooden pen in the shape of a quill*: mahogany, ca. 1875. [9 in. long] and *Carved miniature books* from captured Spanish warship *Reina Mercedes:* mahogany and teak, 1898. *WAF / DB.*

Fig. 7.58 John Haley Bellamy, *American Centennial carved box*: mahogany, 1876. [19½ x 13 x 8¼ in.] *HG / DB*.

Fig. 7.59 Signature of John Haley Bellamy, complete with date. Detail of his *American Centennial carved box*, 1876. *HG / DB*.

Fig. 7.60 Carving of the Great Seal of the United States amidst a vegetal motif. Detail of John Haley Bellamy's *American Centennial carved box*, 1876. *HG / DB*.

Fig. 7.61 Armorial eagle sporting banner and shield. Detail of John Haley Bellamy's *American Centennial carved box*, 1876. *HG / DB*.

Fig. 7.62 Solitary shield. Detail of John Haley Bellamy's *American Centennial carved box*, 1876. *HG / DB.*

Fig. 7.63 Angelic figures. Detail of John Haley Bellamy's *American Centennial carved box*, 1876. *HG / DB.*

this small mahogany box stands as a singular creation, either a commission piece or for Bellamy's own personal use, that serves as one of the most exceptional pieces Bellamy crafted throughout his entire career.

WHILE REVIEWING THE EXTENSIVE DEPTH AND BREADTH of John Haley Bellamy's productive career, it becomes obvious that he stretched his art far beyond the confines of his famous eagles. Rather than being his "refuge," the American eagle was instead the cornerstone upon which he built an amazingly prolific woodworking business. Moreover, it is obvious that his reputation stretched far beyond the confines of Kittery Point's sleepy streets and sun-dappled lanes, casting a large shadow and inspiring many a woodcarver to embrace his example—to the point of outright imitation. His bold new artistic vision had drawn the attention of several artisans eager to follow his lead. Yet it also drew a curious assemblage of friends and admirers, a group that would play a large role in shaping the intellectual identity of Gilded Age America through the written word, the paintbrush, the stage, and in the halls of government and academia—a group that would come to be remembered as "John's fancy friends."

The Lancaster Eagle Part 8

During the many years it spent atop the bow of the *Lancaster* in foreign climes, the Bellamy Eagle's steely gaze, often encrusted with salt spray, beheld five continents and three oceans, innumerable sunrises and sunsets over the globe's waters, and the comings and goings of many exotic people in port. Yet upon her last arrival at the Charlestown Navy Yard, had it pried its eyes open to take in its new surroundings, it would not have found itself in a new museum. Nor would it have spied tourists drawn from near and far to behold it with earnest appreciation. Rather, it would discover that it was still outdoors, baking in the merciless sun from its spot in the middle of the Yard's chain park.

Having been saved from destruction and elevated to the brink of celebrity almost overnight, the *Lancaster* figurehead swiftly descended back down into anonymity and neglect. When plans for Admiral Robinson's Boston Naval Museum fell through, it left the artifact that would have been its star attraction abandoned in a dismal corner of the Yard, exposed to the elements and left to molder.[1] Alternate plans were soon bandied about. Rear Admiral H. A. Wiley assessed the situation hoping to find another suitable home for the titanic sculpture. What he saw did not look promising. A half-century old, the *Lancaster* Eagle was showing the wear and tear sustained over the years; rot had begun to set in. By the estimates of Wiley's office, the cost of providing repairs and an adequate pedestal on which to display the eagle would easily cost over $7,000. Complicating matters was the fact that there was no appropriate location at the Charlestown Yard to display it. Expensive to restore and without a suitable place for display, John Haley Bellamy's masterwork was an onerous burden. The navy's interest in the piece was waning by the day.

Another option soon presented itself. Gordon Dexter, "an influential and public spirited citizen of Boston,"[2] sensing the opportunity to abscond with the magnificent bird at a cheap price, offered to buy the figurehead. Intending to use the eagle as an ornament for his estate north of Boston, Dexter sweetened his offer by promising to display it in a conspicuous fashion upon his property, pledging that the public would be able to view the bird from either the road or from offshore.

As tempting as Mr. Dexter's proposal may have been, the navy's ultimate verdict ignored him. Assistant Secretary of the Navy Theodore Roosevelt (son of the late former president) determined that the *Lancaster* Eagle was to be condemned and sold at public auction. Featured in the *Boston Catalog of Sales of Condemned Material*, the piece first went up for sale on April 20th, 1923, yet no takers were immediately forthcoming. In fact it was not until April 4th, 1925, that the eagle was finally sold for the scanty sum of $262.89 to the Atlantic Marine Exchange of Boston.[3] Located on Boston's Atlantic Avenue, the Atlantic Marine Exchange was a relic from a bygone age when Boston's merchant fleet scoured the seas in the quest for barter and trade and made her a world-renowned capitol of global commerce. The Marine Exchange was a ship's chandlery dedicated to providing the equipment and supplies needed for the outfitting of vessels about to embark upon a voyage, though with naval commerce in Boston being but a whisper of what it had once been the

Fig. 8.1 The USS *Lancaster* Eagle figurehead within the Atlantic Marine Exchange. *Boston Public Library.*

Exchange seemed more like a dusty old junk shop, stuffed to the gills with odd beat-up curios hailing from those long gone glory days.

The decline of the *Lancaster* Eagle from gilded figurehead proudly soaring across the world's oceans, representing the might of the United States, to being perched on the bow of a floating barracks and a quarantine ship, to finally roosting among the secondhand junk of the Atlantic Marine Exchange, [Fig. 8.1] was closely followed by none other than Moses Victor Safford, who lamented the fate of his uncle's handiwork.

> When years afterwards, I saw the eagle a dingy object among the accessories of the head of a disreputable looking housed over old "guardo" with sawed off bowsprit, as the *Lancaster* then was, and when, not many years ago, I saw the eagle hanging up among an assortment of junk in a ship chandler's shop in Boston . . . I felt how inadequate the imagination was to restore the effect which the eagle had as a part of the *Lancaster* as the *Lancaster* looked when she sailed from Portsmouth (MVS, *JHB*, Pg. 19).

Tucked in among the detritus of a once-great port's seafaring heritage, the forlorn *Lancaster* Eagle patiently awaited the day when someone would appreciate its beauty and ransom it from oblivion.

CHAPTER 8

"John's Fancy Friends"

> JOHN HALEY BELLAMY FLICKED A CHIP OFF the half-carved sliver of black walnut. He felt over the dappled wood; then he held it up to the dust-dazzled sunshine which streamed through the southern window of his workshop.
>
> Samuel Clemens (Mark Twain), lolling in a broken chair, pointed his cigar.
>
> "There's what I mean, George." (He was addressing Admiral Dewey.) "See the lift in that wing? As if plucked from an eagle in flight. It's a marvel how he does it."
>
> The admiral nodded. Novelist William Dean Howells smiled assent . . . [1]

With a touch of artistic license did past writers imagine a lazy afternoon in Bellamy's workshop. It may seem more a work of fiction than historic reconstruction, placing important nineteenth-century literary luminaries in Bellamy's humble workshop. Nevertheless, as Bellamy's nephew recalled, these and other American celebrities were indeed regular visitors of Kittery Point's favorite son.[2]

> "John Bellamy's Shop" was a sort of local club. One would find congregated there local master mariners, retired Naval officers and members of the Kittery Point summer colony and their guests. One would often find there men of national prominence. On my visits the notables I recall as having found there enjoying John's ancient but restful cushioned chairs and his company include the Hon. Hamlin Hamlin, Vice-President [during] Lincoln's first term, President Charles Eliot of Harvard, William Dean Howells, Mark Twain and Marion Crawford, the authors, Winslow Homer and George Wasson, the artists, and Edwin Booth, the actor. Henry Clay Barnabee, George Wilson and Tom Burns, the actors of the old Boston Museum company were regular summer habitués of "John's Shop" (MVS: *Eagle*, Pgs. 8–9)

The question naturally arises: how did so many late-nineteenth-century celebrities come to find themselves enjoying John Haley Bellamy's hospitality, and why? One factor that brought these men to Bellamy's door was Kittery's prominence as a vacation resort. Reputed to be a place where one could "enjoy the beaches and the summer breezes,"[3] Kittery proved an irresistible magnet to urban American sophisticates who sought relief from the summer heat while enjoying richly appointed hotels with all the comforts of civilization. Each year, Kittery's fame spread as tourists returned home and extolled not only the wonders of its hotels but the charms of the area and its yearlong residents. For those worldly city dwellers, Kittery and her surrounding environs were a wonderfully preserved vision of an earlier America, untrammeled and unspoiled by the frenetic, fast-paced modern world. The modest, timeworn homes and uncomplicated lifestyles of her inhabitants held a dreamy appeal for those seeking a respite from the age of trolleys and telegraph wires, westward expansion and Gilded Age robber barons.

Themes of rural simplicity and escapism resonated deeply with many of Bellamy's summer guests, and here in Kittery Point they found what they so eagerly sought. A quiet, secluded nook near yet sufficiently removed from the throng of tourists in the area, Kittery Point presented celebrities with the chance to maintain their privacy while enjoying the delights of the Maine coast. Here was a more authentic and intimate window into the everyday people of the area than could be experienced in a hotel, a restaurant, or on a crowded beach. Two American literati who so favored Kittery Point that they purchased summer homes, were William Dean Howells, America's "literary Columbus," who lived down the street from Bellamy, and the artist George Savary Wasson, who lived next door [Fig. 8.2]. Wasson's love for Kittery Point ran so deep that he decided to become a year-round resident.

Whether they sought colorful local characters to inspire their pens, romantic scenes to celebrate upon their canvases, a restful climate, or a nameless, timeless source of cultural renewal where they could reconnect with the early American spirit, Kittery Point grandly fulfilled the needs of these men. It was here that John Haley Bellamy, a native resident of Kittery, and his workshop came to play a central role. Bellamy's workshop was a natural gathering point for various neighborhood characters, including salty old sea captains, hard-bitten fishermen, and veterans of both the US Navy and the Portsmouth Navy Yard. These friends could be found in Bellamy's shop most days, gossiping, swapping stories, and recounting centuries-old bits of local lore—exactly the sort of environment an author or artist on the hunt for the genuine Kittery experience was seeking.[4]

How men like Samuel Clemens (Mark Twain), Winslow Homer, and William Dean Howells first came to hear of Bellamy and make his acquaintance is still unknown. It appears they actually sought him out, for it has been said that in his dealings with summer visitors to his shop, Bellamy "did not force himself upon them. He was never obtrusive. Others had to come to him; those who proved unwelcome were promptly and bluntly made aware of the fact."[5] Safford's memory of a typical afternoon spent in "John's Shop," notes that:

> It was an exclusive club. John attended to that. Nobody succeeded in 'butting in' whose presence John did not welcome. Company in the shop did not interfere with John's work at his bench. Occasionally he might rest himself on a high stool but he kept on working . . . There was plenty of conversation in 'John's Shop' but John took a small part in it, no more than enough to keep it going. He thwarted efforts to draw him into an argument. He was adroit at starting one and enjoyed doing so (MVS: *Eagle*, Pg. 8–9).

Hosting this gathering of great talents was apparently no chore for Bellamy; "He enjoyed the company of such men."[6] And though he generally stood removed from the witty repartee that swirled through his small studio, Bellamy did at times join in on the discussion, entertaining his guests "by relating a humorous anecdote, or by voicing a bit of philosophy that he really believed."[7] Bellamy's pleasure in the camaraderie and friendship of his esteemed circle was not bound to his workshop:

> Occasionally he entertained summer friends by taking them fishing. For this purpose he had acquired an uncouth vessel said to have once been a surf boat somewhere in Central America. Extraordinarily broad for its length, this craft was both staunch and roomy, qualifications that had prompted Bellamy to purchase it. In painting the boat he gave it free reign . . . to his imagination, and achieved something that, under way, made a Polynesian war canoe look like a Sunday School picnic launch. On arrival at the fishing grounds, after furling the sail, he would put up an awning as striking in its color effects as the boat itself. Bellamy's fishing yacht enjoyed the distinction of being the only one with an awning ever seen on the fishing grounds, as well as the only local fishing boat ever known to be equipped with seat cushions.[8]

Among the many who frequented his shop, it was with the artist/author/yachtsman George Savary Wasson [Fig. 8.2] that Bellamy enjoyed the closest and most long-lived friendship. The son of a Unitarian minister, Wasson was born in Groveland, Massachusetts, on August 27, 1855. As a child growing up in Concord, Massachusetts, the future writer first encountered the world of

Fig. 8.2 George Savary Wasson, artist, author, member of the Kittery Point literati, student and close personal friend of John Haley Bellamy: photograph, ca. 1900. *Private collection.*

American literature through his family's acquaintance with both Ralph Waldo Emerson and Henry David Thoreau. (These two relationships inspired his father to take up the pen and write poetry and essays.[9]) He met his second love, the ocean, soon after when his parents moved from Concord to the shipbuilding community of Medford, Massachusetts. Here the young Wasson "learned to handle boats and how boats were built" along the meandering Mystic River.[10] Wasson's love for all things nautical was further strengthened when he spent his summers in West Brookville, Maine, where his grandfather owned a boatyard that repaired and built fully rigged brigs and at least one bark.[11]

Wasson's third love, fine art, was realized in 1872 when the eighteen year old began to study with a Professor Funk for three years in Stuttgart, Germany.[12] The three years he spent in Germany proved quite fruitful for the aspiring painter; he had attained enough skill to merit membership in Boston's Paint and Clay Club and the elite St. Botolph Club. There he came to associate with a number of leading artists, including Daniel Chester French, John Singer Sargent, and Augustus St. Gaudens.[13]

Relying upon his lifelong knowledge of the ocean in its many moods and the vessels that plied their trades across her, Wasson produced numerous romantic seascapes and portraits of inlets and harbors, and made a specialty out of picturesque shipwrecks. His efforts were not in vain. As he "knew every timber in the vessels he drew, the cut a sail should have, and just how the wind from a certain quarter would fill them on a given course," Wasson was soon recognized as "one of New England's foremost marine painters."[14] With a string of exhibitions at the Boston Art Club, the Paint and Clay Club, the Museum of Fine Arts, Boston, J. Eastman Chase's Gallery, Boston, and the Pennsylvania Academy of Fine Arts, the young artist quickly drew an enthusiastic, affluent clientele; it became "the fashion to own at least one of young Wasson's marines."[15]

In 1889 Wasson and his wife Amelia moved to Kittery Point to find an inspiring locale conducive to his art. Building "a house with a studio in the top story, where half-models of ships and nameboards from vessels wrecked along the coast made sympathetic companions for his own paintings," Wasson discovered his home was right next door to the summer home of William Dean Howells and a short distance down the street from the Pepperrell Mansion.[16] Through William Dean Howells, who convinced him to submit stories to the *Atlantic Monthly*, Wasson embarked upon a literary career.[17] Over the years, out of economic necessity, Wasson turned his efforts increasingly toward a writing career, as "the vogue of wall pictures diminished and the merchant marine," the very source of interest in Wasson's paintings, "decreased in numbers."[18] Drawing upon the rich lore and character of Kittery Point, Wasson wrote three books *Home from the Sea*, *Cap'n Simeon's Store*, and *The Green Shay*, works of fiction that established him as the author of "the most authentic Maine stories ever written."[19] His uncanny ability to capture

Fig. 8.3 (left) Attributed to John Haley Bellamy, *Office sign with eagle*: painted pine, ca. 1889. [8¾ x 27 in] *MMM*.

Fig. 8.4 (below) George Savary Wasson, *Copy of a Bellamy Eagle*: stained and painted pine, 1889. [7 x 16 x 4 in.] *Private collection. DB*.

every nuance of the peculiar Kittery dialect (nearly a separate language unto itself) made him "the best dialect writer of America,"[20] and drew the praise of America's master storytellers Bliss Perry, Henry James, Edward Markham, Sarah Orne and Mary Jewett.[21] Even Mark Twain declared after reading Wasson's sketch *Rusticators at the Cove* (featured in *Cap'n Simeon's Store*) that it was "the funniest story I ever read."[22]

Presumably it was Howells who first introduced Wasson into the exclusive circle gathered in Bellamy's shop. We do not know exactly when Wasson made Bellamy's acquaintance, but the two swiftly discovered that they shared a common love: the sea. Both men were yachting aficionados, the products of maritime communities; both had honest, unpretentious dispositions and, as we shall see, a passion for writing. While Bellamy had his "uncouth" surfboat from Central America, Wasson "was exclusively a sailing master," who from the moment he "became a marine painter . . . was never without a boat."[23] One other aspect of the friendship they shared was a mutual interest in the other's chosen profession. Wasson was the obvious source of the artistic instruction by which Bellamy learned how to wield a painter's brush and render rare paintings of the American eagle [Fig. 8.3]. Likewise Wasson learned the woodcarver's trade from Bellamy.

A testimony about how capable Wasson became under Bellamy's tutelage can be seen in his copy of the classic Bellamy eagle executed in February of 1889 [Fig. 8.4]. This eagle in

Fig. 8.5 Inscription upon the back of George Savary Wasson's *Copy of a Bellamy Eagle*, 1889. *Private collection. DB.*

question is nearly identical to Bellamy's handiwork, startling in its exactitude. From Bellamy's trademark 90-degree angle beak (which over the passage of time became fractured and broke off, as we see it today) to the gentle upward curve of the bird's neck, the delicate, free-flowing feather and distinctive *S*-shaped squiggles crowning the top of the raptor's head, the graceful feather cuts, and the dishing running along the top of the eagle's wings, Wasson shows himself intimately familiar with the many nuances of the Bellamy eagle. Perhaps most startling about this sculpture is that it was only Wasson's first such carving; on its back is an inscription in Wasson's hand: "*First Attempt*. Feb. 15, 1889—Kittery Point, ME" [Fig. 8.5].

Thus, by the early months of 1889 Wasson had already won acceptance into Bellamy's exclusive "club." If this was Wasson's first attempt at a Bellamy eagle, then how many other attempts did he whittle? And what compelled him to craft such a bird? If Wasson was carving as a hobby or just out of curiosity, it is likely the number of eagle carvings he made was negligible at best. Or was Wasson not only Bellamy's friend but his employee? We know Bellamy employed both family and assistants to help meet production deadlines. It is also known that Wasson would eventually turn to other pursuits (such as writing novels) to produce an income that offset declining sales of his paintings. With a wife and two small boys to provide for, Wasson may not only have been associating with Bellamy but also working for him when the need arose, helping to carve and paint the large orders that came into Bellamy's shop.

Whether Wasson learned woodcarving from Bellamy as a pleasant diversion or as a way to supplement his income, it is obvious he was the recipient of exceptional schooling. Other carvings by Wasson are no less brilliant in their conception and execution. In the same year as his first eagle, Wasson crafted a striking rendition of the Wasson family crest [Fig. 8.6]. Bellamy's influence is immediately evident throughout this piece, from the subject matter (a heraldic device) to the graceful pair of wings that adorn the shield, even its purpose as a sternboard decoration for Wasson's yacht, the *Bonnie Doon* [Fig. 8.7]. With its flawless symmetry, deft handling, and trademark Bellamy-like handling of the feathers, this piece would, like Wasson's eagle, undoubtedly be mistaken for a Bellamy creation if not for the inscription Wasson scribbled upon its back [Fig. 8.8].

Wasson, like Bellamy, did not content himself with two-foot-long eagles and large plaques. From Bellamy he also learned decorative house carpentry, applying those lessons in his own home. From a lintel over a doorway [Fig. 8.9] to the mantel over a fireplace, [Fig. 8.10] he enlivened his surroundings using knowledge gained firsthand at his friend's side. Wasson's sketchbooks show that Bellamy's love for heraldry had also rubbed off on him; he sketched a picture of the Pepperrell family coat of arms on November 15, 1897 [Fig. 8.11]. And, in a telling clue of just how closely these two friends collaborated in their art, Wasson occasionally painted seascapes upon highly decorative one-piece pine panels that Bellamy carved. [Fig. 8.12] He even painted on panels he carved himself, without assistance from Bellamy [Fig. 8.13].

Fig. 8.6 (top left) George Savary Wasson, *Wasson family coat of arms for the yacht* Bonnie Doon: painted pine, 1889. [16 x 43 x 1¾ in.] *Private collection. DB.*

Fig. 8.7 (bottom left)George Savary Wasson's yacht *Bonnie Doon*, with his Wasson family coat of arms upon the stern: photograph, 1890. *Private collection.*

Fig. 8.8 (bottom right) Inscription upon the back of George Savary Wasson's *Wasson family coat of arms for the yacht* Bonnie Doon, 1889. *Private collection. DB.*

With Wasson, Howells, Homer, Clemens, and others by his side, John Haley Bellamy spent his summer afternoons taking pleasure in a fraternal world of clever debate and earnest discussion, fishing and cruising, and at times instructing in the ways of his craft. The artists, authors, politicians, and actors who befriended him found Bellamy a kindred soul, a gracious, colorful host who entertained them while respecting their boundaries. Like them he was cultured, literate, a touch unconventional, and enjoyed a taste for the finer things in life.

A priceless picture of the artist as his close friends knew him reveals there was more to the man than woodcarving:

> Bellamy had what is commonly referred to as an artistic temperament. He proclaimed this by his fancy waistcoats when fancy waistcoats were in vogue. Out of working hours he dressed well. The artist in him enabled him to appreciate appetizing cookery and all the other good things of life; and he was willing to put money into circulation in order to obtain them. He was not one

Fig. 8.9 (top) Carving upon the lintel of a doorway within the Wasson family's former Kittery Point home. *KC, 2008.*

Fig. 8.10 (bottom) Carving by George Savary Wasson upon a fireplace mantel within the Wasson family's former Kittery Point home. [4¼ x 48 in.] *KC, 2008.*

Fig. 8.11 George Savary Wasson, *Pepperrell Arms*: pencil on paper, November 15, 1897. [6¾ x 4¼ in.] *MSM*.

Fig. 8.12 (top right)John Haley Bellamy and George Savary Wasson, *Near Sea Point, Gerrish Island, Kittery Point, Maine*: oil paint on varnished pine board, ca. 1900. [6 x 8¾ in.] *PMS / RM*.

> of those fabulous beings who are ready to starve themselves in the service of art. No overmastering ambition was ever permitted to interfere with his comfort.[24]

Bellamy's summer friends discovered he was quite unlike the stereotypical, provincial Kittery Yankee—hardnosed, pragmatic, and skeptical of all that was new and unfamiliar—which must initially have taken them by surprise. Bellamy's education "did not stop at school"; throughout his life the woodcarver was "a student of history, literature, and art," as well as heraldry (MVS: *JHB*, Pg. 5). This can be seen in a list of books and their authors written in Bellamy's hand and discovered within his house after his death. Why he scribbled down these titles is unclear, but the list reveals that Bellamy was certainly aware (if not the owner) of fine literature: Nathaniel Hawthorne's *The House of the Seven Gables* and *The French and Italian Notebooks*, Charles Dickens' *David Copperfield*, Thomas Carlyle's *Life of Frederick Schiller* and *Life of John Sterling*, and assorted works by Samuel L. Coleridge are listed. These suggest Bellamy had the working literary vocabulary needed to hold a conversation with the likes of his summer friends.[25]

Fig. 8.13 George Savary Wasson, *Whaleback Light*: oil on varnished pine board, ca. 1889. [12 x 8 in.] *Private collection. KC, 2008.*

Whether it was purely the product of his inventive mind, merely another means for earning a living, or something stirred directly through his associations with Wasson and other noted authors, Bellamy "derived a small income from his writing,"[26] though he never sought the critical acclaim experienced by Clemens, Howells, Crawford, and Wasson. Bellamy nevertheless was a prolific writer in his local community, who contributed "articles on local history and politics and current news, and sometimes poems, to New England papers" (MVS: *JHB*, Pg. 5). He was in fact "a regular correspondent of newspapers" (MVS: *Eagle*, Pg. 9), jotting down his musings and poetic stylings within his cozy bedchamber in the Pepperrell Mansion, seated at a desk that had once belonged to Sir William Pepperrell himself.[27] Given his father's years of governmental experience, it is not surprising to find that politics was a prime subject for his pen. "He was always a devotee of civic change," we are told, with the woodcarver engaging "time and again in fiery conversations and signed letters to the Portsmouth daily newspaper," where "he proposed and sponsored various projects; never to much effect, unfortunately."[28] True to his ability to start an argument, Bellamy's journalism provoked a heated controversy on at least one famous occasion.

> His inclination to provoke discussion once caused him considerable embarrassment. He had written several articles for a newspaper regarding a local project that he strongly favored. Failing to awaken the public interest that he thought the matter deserved, he signed a fictitious name to a letter not only opposing the project but personally attacking John Bellamy. Disguised as his own enemy, he characterized himself as erratic, a crank, and even a good-for-nothing, adding references to incidents purporting to justify such charges. The result exceeded expectations, but not in the manner intended. The project in which Bellamy was interested was promptly obscured in the public mind by indignation at the personal attack. So strong became the feeling against a fellow-townsman wrongly suspected

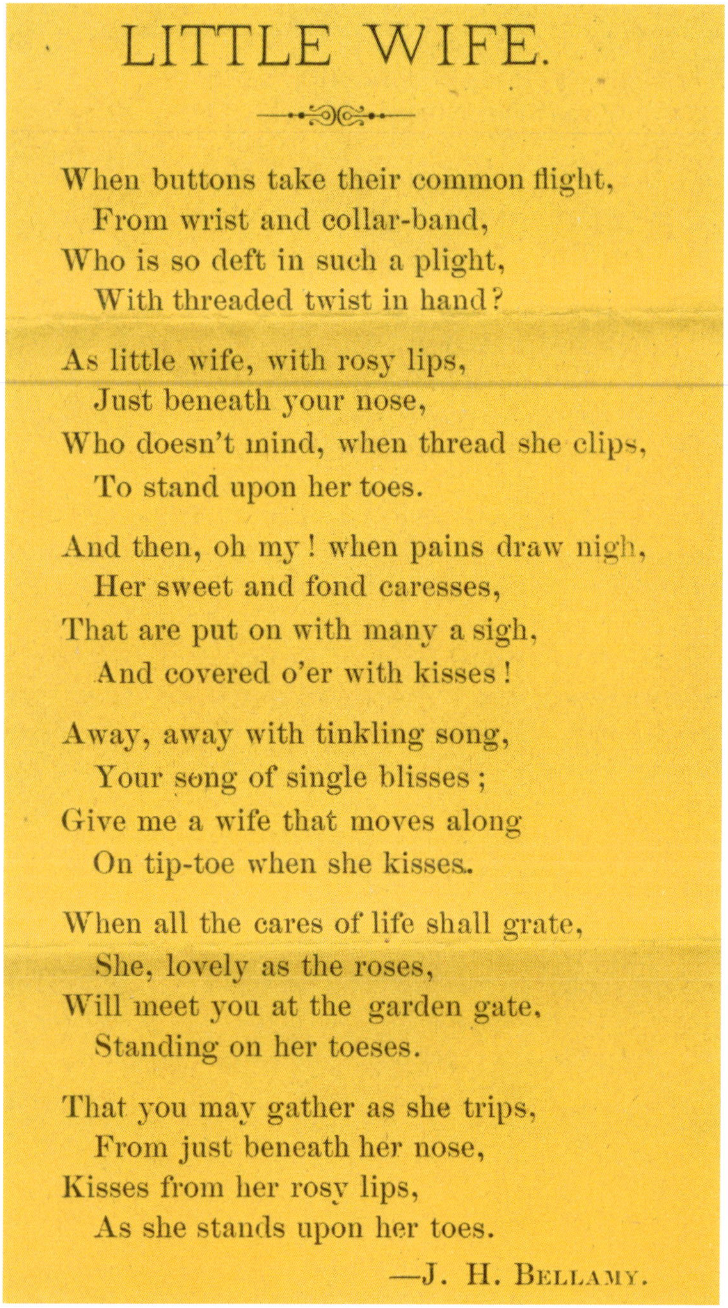

LITTLE WIFE.

When buttons take their common flight,
  From wrist and collar-band,
Who is so deft in such a plight,
  With threaded twist in hand?

As little wife, with rosy lips,
  Just beneath your nose,
Who doesn't mind, when thread she clips,
  To stand upon her toes.

And then, oh my! when pains draw nigh,
  Her sweet and fond caresses,
That are put on with many a sigh,
  And covered o'er with kisses!

Away, away with tinkling song,
  Your song of single blisses;
Give me a wife that moves along
  On tip-toe when she kisses.

When all the cares of life shall grate,
  She, lovely as the roses,
Will meet you at the garden gate,
  Standing on her toeses.

That you may gather as she trips,
  From just beneath her nose,
Kisses from her rosy lips,
  As she stands upon her toes.

—J. H. Bellamy.

Fig. 8.14 *Little Wife*, by John Haley Bellamy. *WAF*.

of writing the abusive letter that Bellamy was obliged to confess his duplicity.[29]

Though his political musings might kindle an occasional firestorm, when Bellamy chose to wax poetic, the result garnered a more serene and pleasing response. Bellamy's poems, a truly neglected, misunderstood facet of his life, demonstrate a healthy imagination and true competency in their handling. Their tone and voice stray from folksy, comical rhymes thick with local flavor to formal Victorian stanzas fit for elevated audiences. That Bellamy took his poems seriously is revealed by multiple drafts of certain pieces, clearly demonstrating that he sought to improve his writings. He even solicited advice from no less a talent than Mark Twain (his friend Samuel Clemens) to improve one unnamed piece, scribbling "Dear Sam Be pleased to fix this over for me. J.H.B."[30] on the back of one poetic draft. And, as others have noted, Bellamy held enough confidence in his skills to publish his poems in both the local papers and as brightly colored handbills.

Unlike his famous woodcarvings, which stand as testimony to his skill as a carver and a businessman, or his political tracts, which reveal his civic concerns, Bellamy's poetry stands as the most revealing window into his thoughts, beliefs, and philosophies, offering us a most intimate view into the soul of the great woodcarver. Free to express his deepest fears and joys upon the blank page, Bellamy repeatedly celebrated themes that held his enduring interest. Surprisingly, one of these was marriage. Though a lifelong bachelor who never tied the knot, Bellamy wrote pieces centered upon marital relationships that covered both its joys and sorrows, such as the charming *Little Wife* [Fig. 8.14] or the reflective verses of:

SO

It is when the young man is thinking of her,
  And when she is thinking of him,
That life is sweetest I vow to you sir!
  And its cup is filled to the brim.

Now what to me is the strangest of things,
That since we know it is so;
We give to our joy the fleetest of wings,
And bid the sweet angel to go.

Yet daily we see the lovers that were,
In marriage, terrible grim;
And all because he's not thinking of her,
And she is not thinking of him.[31]

The inexorable passage of time also gave rise to some of his more inspired works. Within his visionary *To-Morrow,* Bellamy shows a deeply powerful insightfulness, seeing beyond personal ego and the immediacy of his age, realizing the collective mortality of all civilizations as time moves on relentlessly:

TO-MORROW

Onward I move amid the throng
Who jostle on this rolling sphere,
And feebly murmur, this my song,
We all shall shortly disappear.

A little month, a passing year,
And we have reached life's journey's end.
Beyond the fount of grief and tear,
The land of promise, home and friend.

Still there will float in upper air,
The storm cloud on its trackless way,
And fawning doves will coo and pair,
As fondly as they do to-day.

There will the unborn millions rear,
Their temples high throughout the land,
While crumbling, ours shall disappear,
Beneath Time's all destructive hand.

Heroes will lead in unknown deeds,
And ships make havoc on the deep,
New kings expound and break their creed,
New babies in their cradles sleep.

Thus runs the busy world away,
Without the breaking of that law,
That hath the universal sway
Dealt with our kings as men of straw,

'Tis well methinks there is a power.
Beyond the reach of human greed,
That nameth not the day or hour,
But bids us all alike take heed.[32]

Poems about toothaches, the dawn of springtime, and Kittery's tourism industry flowed from Bellamy's glib pen now and again, prosaic topics of a simpler, more modest nature. Yet for every poem he wrote about marriage, the passing of civilizations, and the everyday joys and trials of life, the woodcarver's poetry centered upon a deep, at times even morbid, fascination with his mortality. While he had no death wish, Bellamy certainly reflected upon his mortal passing on a regular basis, repeatedly voicing a wistful imagining of his burial in works like his *Bury Me Beneath the Rock* and *Cenotaph*. [Fig. 8.15] Others disclose his later years had, despite his famous friends and successful business, become a chore. As he moved beyond middle age into his elder years, the artist found in poetry an ability to voice his fears and angst, while praying that in his remaining years he could be spared from the pains of aging. Even in the midst of such cheerless reverie, Bellamy could still fire off a frank, delightfully vulnerable and immediately human commentary on his advancing years, free from melancholic Victorian sighs and drama:

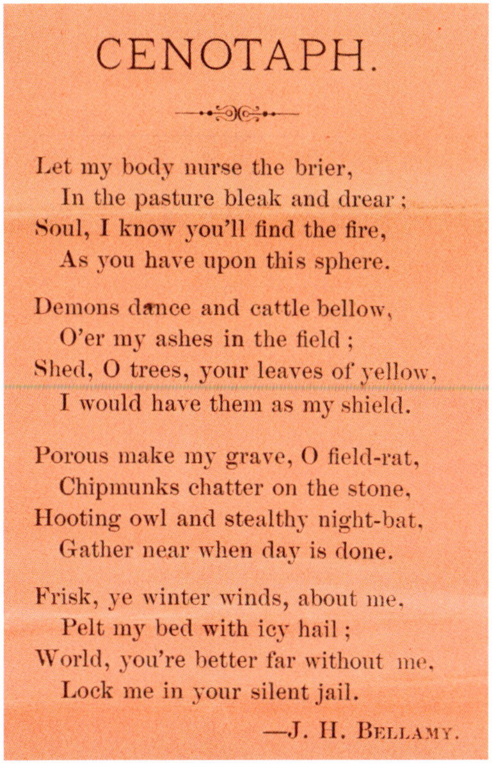

CENOTAPH.

Let my body nurse the brier,
In the pasture bleak and drear;
Soul, I know you'll find the fire,
As you have upon this sphere.

Demons dance and cattle bellow,
O'er my ashes in the field;
Shed, O trees, your leaves of yellow,
I would have them as my shield.

Porous make my grave, O field-rat,
Chipmunks chatter on the stone,
Hooting owl and stealthy night-bat,
Gather near when day is done.

Frisk, ye winter winds, about me,
Pelt my bed with icy hail;
World, you're better far without me,
Lock me in your silent jail.

—J. H. Bellamy.

Fig. 8.15 *Cenotaph*, by John Haley Bellamy. *WAF*. This and other poems were printed upon a letter press within the Pepperrell Mansion. The reasons for their printing remain unclear.

GRANDPA

Harder do I daily lean
Upon my oaken staff,
And as I plod my weary way,
The merry children laugh.

Once I was young and nimble too,
But now, alas, alas,
Time's cruel hand has crippled me,
Young villains, let me pass.

Fig. 8.16 John Haley Bellamy in his mid-sixties: photograph, circa 1900. *WAF*.

O'er across yon winding mead,
Where birches are so tall,
I romped, like you, from school tasks freed,
The merriest boy of all.

But now I am aweary, child,
It's the long race I've run,
Youth's heady sports and gambols wild,
Are mine no longer, son.

Soon I shall pass unto my rest,
While you remain to play;
So spare the cruel jeer and jest,
While grandpa goes his way.[33]

A photo taken of Bellamy about 1906, [Fig. 8.16] captures a hint of what life was then like for the once dashing, hale artist and why his focus on aging and death. As related by his nephew:

Bellamy must have been nearly if not quite seventy years old when the snap-shot was taken although to one who did not know him in his prime he would not appear so old. He is wearing his workshop clothes and carver's apron and was evidently snapped while walking to or from his workshop and the Post Office at Kittery Point. One familiar with his younger erect dignified carriage will see in his slumped figure a betrayal of the arthritic trouble with his feet which bothered him in his late years. The shadow of his hat distorts somewhat his features and his Roman nose but on the other hand it is an unposed natural photograph and shows him as we knew him, and in motion, and best of all he was "snapped" as he turned to see what the photographer was doing and the photographer caught the quizzical smile which we have seen him show when somebody tried to start an argument in his "Shop" (MVS: *Eagle*, Pg. 17).

For Bellamy, time bore down upon him more and more as the nineteenth century drew to its close. Riddled with arthritis, he could no longer deny his mounting age. The final decades of the century were still passing pleasantly enough. He still had an active social life, surrounded by celebrated friends, and was continuously busy with his successful woodcarving career. And the inventive and entrepreneurial side of Bellamy was still afire. As late as November 26, 1900, at the age of sixty-four, he was still contriving improved tools and devices for the Massachusetts Humane Society, a private organization that served to assist and or rescue those in distress on the high seas, long before the US Coast Guard took on the job.[34] Around the year 1895, in a move designed to capitalize on the enormous attention that was then being generated by the festivities surrounding the one-hundred-and-fiftieth anniversary of Sir William Pepperrell's capture of the French fortress of Louisburg in 1745, Bellamy and his family opened their historic home as a tourist attraction, going so far as to print up numerous handbills advertising the home's historic significance and charms [Fig. 8.17]. The handbill, written by a family of woodcarvers, placed a premium upon the quality

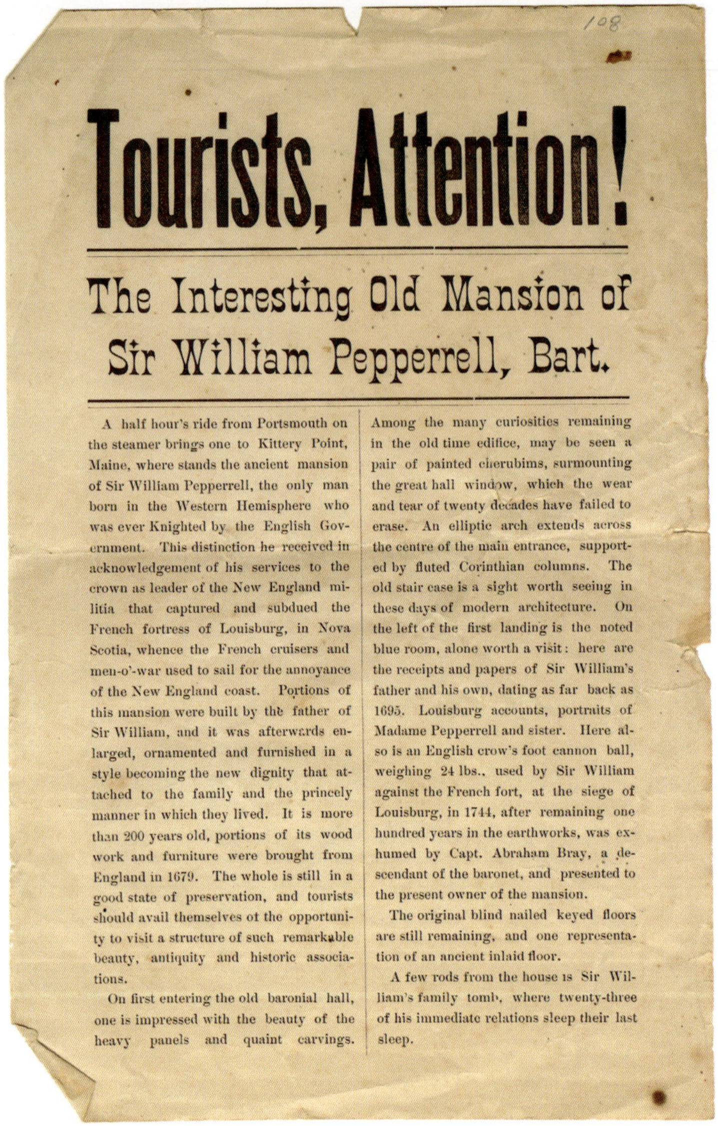

Tourists, Attention!

The Interesting Old Mansion of Sir William Pepperrell, Bart.

A half hour's ride from Portsmouth on the steamer brings one to Kittery Point, Maine, where stands the ancient mansion of Sir William Pepperrell, the only man born in the Western Hemisphere who was ever Knighted by the English Government. This distinction he received in acknowledgement of his services to the crown as leader of the New England militia that captured and subdued the French fortress of Louisburg, in Nova Scotia, whence the French cruisers and men-o'-war used to sail for the annoyance of the New England coast. Portions of this mansion were built by the father of Sir William, and it was afterwards enlarged, ornamented and furnished in a style becoming the new dignity that attached to the family and the princely manner in which they lived. It is more than 200 years old, portions of its wood work and furniture were brought from England in 1679. The whole is still in a good state of preservation, and tourists should avail themselves ot the opportunity to visit a structure of such remarkable beauty, antiquity and historic associations.

On first entering the old baronial hall, one is impressed with the beauty of the heavy panels and quaint carvings. Among the many curiosities remaining in the old time edifice, may be seen a pair of painted cherubims, surmounting the great hall window, which the wear and tear of twenty decades have failed to erase. An elliptic arch extends across the centre of the main entrance, supported by fluted Corinthian columns. The old stair case is a sight worth seeing in these days of modern architecture. On the left of the first landing is the noted blue room, alone worth a visit: here are the receipts and papers of Sir William's father and his own, dating as far back as 1695. Louisburg accounts, portraits of Madame Pepperrell and sister. Here also is an English crow's foot cannon ball, weighing 24 lbs., used by Sir William against the French fort, at the siege of Louisburg, in 1744, after remaining one hundred years in the earthworks, was exhumed by Capt. Abraham Bray, a descendant of the baronet, and presented to the present owner of the mansion.

The original blind nailed keyed floors are still remaining, and one representation of an ancient inlaid floor.

A few rods from the house is Sir William's family tomb, where twenty-three of his immediate relations sleep their last sleep.

Fig. 8.17 *Tourists, Attention!* Handbill advertising the Pepperell Mansion as a tourist attraction. *WAF*.

of the decorative woodwork found in the Pepperrell Mansion: "On first entering the old baronial hall, one is impressed with the beauty of the heavy panels and quaint carvings"; "An elliptic arch extends across the centre of the main entrance, supported by fluted Corinthian columns"; and "The original blind nailed keyed floors are still remaining."[35]

Yet still the sense of human mortality and frailty remained, drawing closer with every breath. No matter how many carefree days he spent in the company of friends and family, no matter how busy and preoccupied he was with work, Bellamy's poetry increasingly centered upon thoughts of a better life beyond the mortal realm. And though hidden beneath obscure mythological references, a sincere desire for release began to appear, as in his poem *Atropos*. Addressed to the ancient Greek goddess of Fate who determined when every life was destined to expire by severing the thread of life on the spindle around which it was wound, Bellamy's poem is in fact an oblique prayer for death's release.

ATROPOS

Atropos, Dear, my thread's too long
    O break it on the spindle
Alluring fires lead me wrong
    The fires demons kindle—

When hot about my flanks—I hop!
    It won't be counted, will it?
No more'n the corn that lovers pop
    At midnight in the skillet—

Oft has it led my weary legs
    Through tangled brakes and mire
Where wild birds hover suckled eggs
    And bounteous is the brier

Against the running curent [*sic*] too
    Oft have I been headed
While round about the chosen few
    In comfort are well beded [*sic*]

So strike I say and break the tie
    Thus end my world of sorrow
And send me to the by an by
    Sweet home of the tomorrow[36]

Even as a new century was unfolding, the life of America's greatest woodcarver was beginning to fade into the twilight. Bellamy's thinly veiled prayer had been heard. Atropos would soon strike.

The Lancaster Eagle Part 9

For nearly a decade the *Lancaster* Eagle figurehead languished within the dim, dusty precincts of Boston's Atlantic Marine Exchange, a forlorn curio in a room full of maritime novelties. Having acquired the raptor for only $262.89, the owners of the Exchange now sought to profit handsomely from their purchase by asking no less than $3,500 for the piece, a high price that, coupled with the inflexibility of its new owners to negotiate and the sheer size of the figurehead, foiled more than a few attempts to liberate it. Interested parties had enquired from time to time about the mighty eagle: one local manufacturer is reported to have tried to buy the bird with $1,500 cash and $1,000 in bank notes, only to be informed that the Exchange would only accept cash. Another firm is rumored to have considered making a copy of the majestic artifact as a less expensive alternative, yet abandoned the notion when discovering that having a copy made would cost even more.[1]

Deliverance for the *Lancaster* figurehead ultimately came in January 1934, in the form of an antiques dealer from Westport, Connecticut, by the name of Earl Smith. Acting as an agent for the newly formed Mariner's Museum of Newport News, Virginia, Smith had come to Boston specifically to purchase the great sculpture, something he soon discovered was no easy feat. Despite the fact that the nation was thoroughly caught in the financial morass of the Great Depression and offers to purchase the Eagle were few, if any, the proprietors of the Exchange were still averse to selling their prize for anything less than $2,500. For Earl Smith, who had a check from the Mariner's Museum for only $1,500 in hand, the response to his initial overture was most disheartening. A bout of haggling ensued, with Smith purportedly declaring that he had heard of a written offer by the Exchange to sell the eagle for only $2,000, to which one of the obstinate owners is said to have replied, "Oh yeah, well you show me where I have offered it for less than $2,500 and I'll give it to you."[2]

Round one of the negotiations had gone poorly for Smith and the Mariner's Museum. Matters were further complicated when the museum's manager William Gatewood suddenly had a change of heart during the negotiations; he fired off a letter to Homer L. Ferguson, president of the Mariner's Museum, declaring, "I recommend not purchasing. It would be too difficult to display."[3] Homer Ferguson would not be so easily dissuaded, however, and so began a second round of discussions, complete with a new overture from the new museum in Virginia. For reasons never clearly defined, the stubborn, opportunistic owners of Bellamy's masterpiece consented to the new terms and with a check for $2,200 and a handshake, the Mariner's Museum became the new owner of the figurehead in late January 1934. After arduous preparation to assure its safe arrival, by February of that year the *Lancaster* Eagle found itself migrating south to a warmer clime, destined to finally take its place of honor in a maritime museum.

Upon arrival, the first task for the museum was creating a suitable display area for the sculpture, something much easier said than done given its enormous size and weight. In the simplest and most expedient means then available, the Mariner's Museum elected to station Bellamy's bird outside on a concrete pedestal, with guy wires wrapped around its throat to secure it to an exterior wall [Fig. 9.1]. Intended from the start as a temporary solution, the figurehead would wait patiently upon its narrow stoop outside until 1936 when, with the construction of its new quarters finally completed, it was moved inside to assume its proper place as the centerpiece of the museum's main gallery.[4]

In an attempt to hide the scars of time and deterioration, the Mariner's Museum elected to repaint Bellamy's masterpiece; it had already been painted at some point between its debut in August 1881 and its purchase in 1934. Cloaked in hues of chocolate brown with white accents, the bird looked almost absurd. Yet even so, the majestic sweep of its neck, its soaring wings, and sheer size easily impressed the new museum's visitors and patrons. With each year, the fame of Bellamy's masterpiece spread, and with it so did the reputation of the institution that so proudly displayed it. Eagle and museum became inextricably entwined, each becoming synonymous with the other as they sparked a sincere academic interest in the creator of the noble figurehead. By the late 1950s/early 1960s, the Mariner's Museum took it upon itself to properly restore the artifact that had evolved into its trademark, an operation that demanded the removal of the offending paint that marred it. Scraped down to the bare wood, the piece was then painstakingly regilded, [Fig. 9.2] an arduous and expensive process that when completed finally returned it to its full, uncompromising grandeur.

In the years since its regilding, the *Lancaster* Eagle has grown in eminence and importance. Today it stands as the unrivaled example of the American nautical woodcarving tradition at its finest, representing the perfect marriage between aesthetic principles and sound engineering precepts. Likewise, it has become the enduring symbol of one of the world's truly supreme maritime museums. No longer residing within one of the Mariner's Museum's galleries, it stands today within that institution's entrance hall, welcoming innumerable visitors from shores near and far, testifying to all the world the strength and greatness of America's maritime heritage.

Fig. 9.1 The USS *Lancaster* Eagle figurehead on temporary display outside its new home, The Mariner's Museum, Newport News, Virginia: photograph, 1934. *MM.*

Fig. 9.2 The USS *Lancaster* Eagle figurehead during its regilding: photograph, early 1960s. *MM.*

Fig. 9.3 The last known photograph of John Haley Bellamy, taken in his carving shop: photograph, ca. 1910. *WAF*.

CHAPTER 9

A Life in Decline

WITH THE DAWN OF THE TWENTIETH century, John Haley Bellamy, woodcarver, inventor, poet, and friend to several literati, found himself drifting into waters he had never before traversed. An industrious soul who had never been far from the warm affections of family and friends, Bellamy's world was becoming one of increasing solitude and lassitude:

> In his closing years Bellamy led what has been referred to as a 'secluded life': 'lonely' would be a more appropriate term. He still kept mentally occupied and informed regarding current events. He continued to be interesting company; but he reflected loneliness. With the husband of a dead sister and her unmarried son, he maintained a bachelor ménage in the old home. A housekeeper looked after the trio. John was the last survivor of eleven children. Almost empty of life, the great house was yet crowded with memories: memories of familiar faces that had gone, of the prosperous days when Bellamy's father and mother, famed for their hospitality, had held high entertainment and the mansion had resounded with the voices of their brothers and sisters, and of a jovial company of guests; memories, too, of receding fortunes, of the succession of deaths that had bereft the ageing man of his nearest and dearest kin. Not happy memories, these.[1]

With each passing year, the Bellamy family had inevitably dwindled in number. Bellamy's sister Martha Ann was the first to die in 1880.[2] She was followed by Bellamy's brother Elisha in 1884 (the tragic victim of a train strike in the town of Malden, Massachusetts), sisters Frances in 1887, Laura in 1897, and Sophronia in 1899.[3] Frances Keene Bellamy, the family matriarch, died in 1892, while her husband, the esteemed Charles Gerrish Bellamy, passed away on Friday, January 13, 1899, at 3:30 in the morning, due to a "sudden attack of the grippe."[4] At the turn of the century the once-bustling Pepperrell Mansion had only four tenants dwelling within it: John Haley Bellamy and his brother Charles Carroll Bellamy, who were joined by Zachariah and Eugene Williams, the husband and son of the recently deceased Sophronia Bellamy Williams, their sister. By 1910 the house had further dwindled to only the two Bellamy brothers; the Williams' having moved next door after a familial squabble over ownership of the seaside mansion.[5] The following year Bellamy lost his brother Charles and his sister Josephine, leaving him the last surviving member of the large, prosperous clan.[6]

Living in the increasingly empty shell of a once-bustling household, Bellamy's thoughts during the 1890s and early 1900s increasingly turned inward as he roamed through the vacant chambers of the Pepperrell Mansion, dreaming of former glories and golden ages long past as suggested in his haunting, melancholic poem "Yes."

YES

Here it was Sir William laughed
    And buckled on his shoe
Here it was he often quaffed
    And Parson Newell too.

But since that day, that happy day,
To dust he's gone as fuel;
There's no Sir Will, no Margery Bray,
And no more Parson Newell.

And, 'neath the gable rafters high,
Where dwelt his trusty slaves,
The spider battles with the fly
In rooms, now empty caves.

Than in the cellar, deep and dark,
Where choicest wines were stored,
I feel the thrill of Bacchus' spark,
But where, oh where's his board?

Again! Through empty spacious hall,
I go my thoughtful way;
And gaze for pictures on the wall,
Of lovely Margery Bray.[7]

As family members gradually fell away with the passing years, so did Bellamy's rich cadre of friends and acquaintances:

> Time had also taken toll of Bellamy's cronies. One after another the ancient mariners had dropped off. Many other friends, including George Wasson, had moved away. The character of the summer colony was changing. The newcomers had not the same interests and diversions as their predecessors. The gatherings at John's shop dwindled, until in winter the only daily visitor was the retired shipmaster, Captain John Lawrence. He and Bellamy had been close friends from boyhood, ever since a local shipmaster had brought Lawrence to Kittery Point as an orphan boy, survivor of an emigrant ship tragedy.
>
> One winter day news came to Bellamy in his shop that Lawrence too was gone. For a time the old woodcarver sat quietly on the high stool by his workbench and let his vision rove over the sea to the meeting line of sky and water. Then he arose, left the shop, locked the door, and walked slowly to his room in the old mansion. Never again did he return to the shop. He insisted that he "would go back sometime," but the sometime was postponed. For a year or two he occupied himself with newspapers and periodicals and books in his living room in the mansion. But the love of life had passed from him.[8]

The day Captain Lawrence died was either on Thursday, January 3, or Friday, January 4, 1907. [9] Traditionally this has been held as the exact moment when Bellamy quit carving. However, the handiwork of John Haley Bellamy itself contradicts this idea. Inside a trunk carved by Bellamy, is a most revealing clue that he was in fact carving until his final days {Fig. 9.4]. Decorated with the Bellamy family coat of arms front and center upon its front panel, [Fig. 9.5] the trunk is not as richly decorated or as imaginative as the small mahogany box discussed in chapter seven. Yet the chest clearly reveals Bellamy's unique handiwork in the deft craftsmanship evident throughout. It also informs us that Bellamy was actively carving for at least five years after the passing of his good friend Captain Lawrence, for upon the inside of the trunk's lid can be found the carved words "M. Victor Safford. 1912."

This name and date prominently inscribed upon the chest shows that Bellamy was still carving in a competent fashion two years before his death. So, what led Moses Victor Safford to claim that his uncle stopped working in 1907, when this trunk bears Safford's name and clearly states otherwise? There is no evidence that Safford was trained in woodcarving. The intense similarities between this piece and the known Bellamy carvings are just too close to be copy-work by another woodcarver, even George Savary Wasson. The most satisfactory answer to this mystery is the provenance of the trunk. Uncovered in 1944 by the Kittery Point antiquarian Joseph W. P. Frost when his family took over the Pepperrell Mansion, the chest, along with a horde of Bellamy sculptures, letters, documents, and other ephemera related to the woodcarver, had never left Bellamy's workshop, When Frost first entered it, the shop was "much as he [Bellamy]

Fig. 9.4 John Haley Bellamy, *Carved chest with Bellamy family coat of arms*: mahogany and pine, 1912. [21½ x 40 x 20 in.] *Private collection. DB.*

Fig. 9.5 Carving of the Bellamy family coat of arms amidst a vegetal motif. Detail of John Haley Bellamy's *Carved chest with Bellamy family coat of arms*: mahogany and pine, 1912. *Private collection. DB.*

had left it, except for the dust and decay of thirty years."[10] Perhaps the chest was crafted by Bellamy for his nephew as a gift, without Safford having ever been aware of its existence.

While the chest reveals that Bellamy was in fact still carving beyond the death of Captain Lawrence in 1907, it does not mean that Bellamy was carving regularly until 1912. Bellamy's days of filling orders for several hundred eagles at a time were past, and the carver produced such items only on rare occasions. Old age and infirmities had caught up with him, as evidenced by minor defects in the chest's design, such as the slight irregularities to the ellipse bordering the trunk's brass handle [Fig. 9.6].

Bellamy may have begun to scale down his production at the dawn of the twentieth century. According to the US Navy's rather spotty civilian personnel records, Bellamy is last known as having worked at the Portsmouth Navy Yard during a period stretching from May 22, 1897, to June 19, 1897, when he was discharged due to the ever-chronic "Lack of work."[11] He was officially dropped from the Yard's register of employees on June 1, 1898.[12] Safford, however, would claim, "After 1900 he was still turning out carvings for the Navy at the Portsmouth yard" (MVS: *JHB*, Pg. 11), thus offering a tantalizing (but unverified) possibility of his having remained with the Navy even longer.

Regardless of when he ultimately quit the Portsmouth Navy Yard, by the early years of the twentieth century the woodcarver was no longer deriving a meaningful income from his art. Bellamy was by then living completely off of his past earnings. For three years around 1909, he also collected (but did not spend) an annuity payment for the sum of $750 a year under policy No. 1500,786 from the Equitable Life Assurance Company of the United States, paid out

Fig. 9.6 Detail of John Haley Bellamy's *Carved chest with Bellamy family coat of arms*: mahogany and pine, 1912. *Private collection. DB.*

four times a year in increments of $187.50.[13] Safford claimed, that as Bellamy "would have to be taken care of in his old age led my mother's sister to endow him with an annuity which gave him a comfortable income through life" (MVS: *JHB*, Pg. 4).

While Safford clearly believed this tale, two surviving letters from The Equitable Life Assurance Society of the United States to John Haley Bellamy instead declare that the payments to Bellamy were being made specifically "to your order."[14] All the evidence indicates that Bellamy was drawing money from a policy he had established on his own volition from his own funds. Contrary to Safford's claim that the annuity "gave him a comfortable income through life" however, the annuity was paid out to Bellamy only for a period of three years.

Increasingly bereft of friends and siblings and forced by age and arthritis into a state of general retirement, Bellamy's last years held little to engage him. He "occupied himself with newspapers and periodicals and books" during this time. One of these was the *Atlantic Monthly*; a subscription was taken out in his name by his friend George Wasson, who then lived in Bangor, Maine, but still held an active interest in his old friend's welfare.[15]

Among the hazy, listless days and lonely nights, when Bellamy found himself with little to do and even less to look forward to, the most persistent and infamous of legends concerning the life of John Haley Bellamy was spawned. In the words of one erstwhile Bellamy biographer:

> Life refused to leave the tired spirit of John Bellamy. He lived in a vacuum of his own private hell. He was the town drunk. On warm days, John sat under a great tree across the road, in front of the post office. He became an object of ridicule as neighbors passed on their way to Frisbee's Market.[16]

This picture of John Haley Bellamy's alcoholism has proven quite enduring over the years. Yet in a fashion no different from so many other Bellamy legends, the reality must be carefully separated from the theatrical and the apocryphal. It is true that Bellamy did indeed have a troubled, at times stormy, relationship with intoxicating beverages for much of his life. His mother wrote to her husband in an undated letter around 1870 that, "John has maid [*sic*] no sails [*sic*] since you were down got lot of Egles [*sic*] maid [*sic*] has not drinkt [*sic*] any liqure [*sic*] since you were hear [*sic*] . . ."[17] In his own poetry Bellamy further drops possible hints at his dalliances with drink. In his poem *Atropos*, he declares "Alluring fires lead me wrong / The fire's demon's kindle." In *Yes*, he speaks of feeling "the thrill of Bacchus' spark" in Sir William Pepperrell's former wine cellar. Verification that Bellamy was indeed prone to overindulging in liquor can be found in the police log books of Portsmouth, New Hampshire, where we discover that on October 30, 1891, and again on December 11, 1902, Bellamy was arrested in Portsmouth for the crime of public drunkenness.[18]

Nevertheless, before leaping to the conclusion that Bellamy "lived in a vacuum of his own private hell" as "the town drunk," new evidence must be taken into account. First is the nature of communication within the Bellamy family. Of the many surviving letters in public repositories today related to the Bellamys, the above sentence penned by Fanny to her husband Charles stands as the one statement suggesting John Haley Bellamy had a problem with alcohol.[19] It stands to reason that many more letters like this would surely exist if Bellamy's weakness for drinking was as dramatic and debilitating as previous authors have suggested, especially when those authors claim that by "the summer of 1871, John's drinking had become a serious problem."[20] Yet whether frankly discussing the weather, village gossip, business ventures, or the daily humdrum activities of life in the Pepperrell Mansion, the Bellamy clan was uncharacteristically silent concerning this matter. With their confiding words about personal abilities, weaknesses, hopes, and fears for their immediate kin and neighbors, each of the Bellamys write with honest expression. Secrets among them seem to have been rare, and given their close living quarters it would have been nigh impossible to keep any for long. It is hard to imagine that an enormous gulf of silence about a problem as supposedly public as John Haley Bellamy's alcoholism could have spanned several decades of the Bellamy family's communications. Rather, it would appear to indicate that Bellamy's issues with the bottle were far less dramatic and far more sporadic than sometimes claimed.

That drunkenness manifested itself infrequently is also confirmed by the Portsmouth, New Hampshire police department. Studying the Portsmouth police logs from August 16, 1876 (the earliest known pertinent log book from the Portsmouth police department still in existence), straight through to his death in 1914, Bellamy's name appears only twice. These two instances, separated by more than eleven years, hardly support claims of Bellamy as an inveterate alcoholic. In the eyes of the Portsmouth police department, he would have been a general unknown, especially compared to other Kittery residents whose names appear far more frequently—sometimes weekly. Individuals like Mr. Charles F. Manson, Archie Boardman, and a Mary Taylor, three people arrested for public drunkenness in Portsmouth at almost regular intervals from month to month, were far better acquainted with the Portsmouth police than Bellamy ever was. A particular favorite of the Portsmouth police, Kittery's Solomon Kimball, was arrested many times a year for this crime for well over two decades.[21]

Unlike Charles F. Manson, Archie Boardman, Mary Taylor, Solomon Kimball, and others, Bellamy was not always possessed by an all-consuming thirst for libations. Further indication of this may be deduced from his long-lived association with the Portsmouth Navy Yard from the Civil War until at least 1897. When appraising his "character of conduct and work," his navy yard supervisors described both as excellent, never once making mention of a drinking problem.[22] Indeed, the very fact that Bellamy was employed by the navy for so many years, keeping his job during times of scant work and earning the respect and trust of his military and civilian bosses alike, stands as an impressive endorsement of not only his competency and professionalism on the job but his sobriety as well. Had Bellamy totally lacked self-control as legend claims, he doubtless would have been fired from his navy yard job in a fashion no different than so many other employees found drunk during working hours.

The many years of carvings produced by Bellamy in his Charlestown and Kittery workshops further suggests sobriety having been his usual state. Even with the aid of assistants and industrial production methods, the consistent quality of his carvings, combined with his ability to take on large orders and meet production deadlines, may have been well impossible had Bellamy been habitually inebriated. The report R. G. Dun & Company wrote about J. H. Bellamy & Company in 1869 describes him as "steady and Indus[trious]"; there is not even a whisper of him being an alcoholic, a detail that would not have escaped mention, as evidenced by the sheer number of times it was often noted in the mercantile agents' reports about other businessmen.

It is also important to note that Moses Victor Safford, Bellamy's first biographer, never once felt the need to mention drinking in all his epistles and articles concerning his famous uncle, again suggesting that it was not a source of persistent troubles and embarrassment. Rather, Safford commented, "Unlike some artistic geniuses, he [Bellamy] cherished strict moral principles to which he adhered,"[23] words that, though by no means definitive, certainly allude to Bellamy being far less inclined toward "dissipation" than we have previously been led to believe. That others shared Safford's and the navy's appraisals of Bellamy's moral character is revealed in the following letter from the New York Life Insurance Company:

> We desire to obtain a representative for our Co. in Kittery Point and vicinity, as we know a fair amount of insurance can be written there, and your name has been suggested to us as a man who could undoubtedly [*sic*] sell insurance . . .
>
> If you are situated so that you would not care to devote your entire time to same, we would be pleased to make arrangements whereby you could devote what time you had to spare, and we feel sure that you could materially increase your income from same.[24]

It is hard to imagine a life insurance company considering a highly visible town drunk as their sales representative. It is also difficult to imagine men like William Dean Howells, Samuel Clemens, Winslow Homer, George Savary Wasson, the Honorable Hamlin Hamlin, Harvard College president Charles Eliot, and others regularly associating with a drunkard. While Bellamy and his friends undoubtedly quaffed a drink from time to time, especially when cruising around in his boat, there is no reason to imagine them congregating around a man who could regularly be found passed out drunk "under a great tree across the road" from the Pepperrell Mansion.

The final factor in our appraisal of Bellamy's involvement with alcohol is his association with a particular organization called the Knights of the Maccabees. A nineteenth-century "legal reserve society," the Knights of the Maccabees "were one of the more successful of fraternal benefit societies which sprung up after the Civil War."[25] Formed at a time when many insurance companies did not pursue sales to everyday people of ordinary means, groups like the Maccabees provided a financial safety net along with a venue for pleasant social meetings for their members. While "all white persons of sound health and good character, from birth to 70 years of age, were eligible for membership," certain professions were excluded due to their inherent dangers or promotion of unhealthy lifestyles. This policy expressly included the "Manufacturers, sellers, and drinkers of alcohol."[26] John Haley Bellamy was a "Member of North Star tent #193, located at Kittery Point, ME." having been ordained "Sir Knight John H. Bellamy," on May 20th, 1880. Kittery Point's great woodcarver was a member of an organization directly opposed to the very behavior he has been accused of regularly indulging in.

When pondering the facts at hand, it becomes evident that Bellamy's relationship with liquor has, like so many other aspects of his life, become horribly misconstrued over the passage of time. Yet it is undeniable that Bellamy did, at least on occasion, encounter difficulties with his consumption of alcohol. Though only arrested twice for public drunkenness, the fact remains that he was arrested. And though his family makes scant mention of alcohol as being problematic for him, his mother did once allude to it. From the available evidence, the image of Bellamy that emerges is not that of a hopeless, perpetually intoxicated "object of ridicule." Nor is it one of a chaste, temperate soul prudishly upturning his nose at the mere thought of a drink. Instead we see a man who struggled with—and by and large overcame—his penchant for alcohol. The presence of a printed flyer published by the Washingtonian Inebriates among Bellamy's personal papers [Fig. 9.7] strongly intimates to his having taken an active interest in curbing his alcoholism. This organization was the short-lived forerunner to

Fig. 9.7 Printed flyer for the Washington Inebriates in the papers of John Haley Bellamy. *WAF.*

today's Alcoholics Anonymous, dedicated to "helping the individual alcoholic" through a doctrine of total abstinence from liquor and the creation of "reformatory homes" where alcoholics could receive treatment and support as they wrangled with their addiction.[27]

It is very telling that Bellamy was only arrested for public drunkenness in the last decades of his life and only rarely then. For the vast majority of his years, Bellamy walked a tightrope every day between responsibility and chaos. For the most part it was a tightrope he walked successfully. Had he failed, his artistic creations would have been second-rate offerings at best, the undisciplined hackings of yet another backwoods whittler, instead of the masterworks we can hold in our hands and marvel at today. Had he indeed succumbed to becoming Kittery's town drunk, he never would have been able to produce the extraordinary number of diverse, imaginative sculptures that flowed out of his shop, and consequently we would not be reading about him today. Bellamy would have passed from this earth in anonymity, quickly forgotten and unlamented. Instead he is remembered as nineteenth-century America's most inventive and original woodcarvers, the herald of a new form of artistic expression, the creator of his country's finest figurehead. In his struggle with the bottle we see Bellamy at his most human and vulnerable—and at his most triumphant.

As to where and how Bellamy's name came to be synonymous with alcohol, the words of Horace Mitchell, a former writer on Bellamy who, like Safford, was generally more judicious in his research than past authors and less prone toward sensationalism, succinctly sums up this phenomenon:

> Based on one or two truths, these yarns, now more than fifty years after Bellamy's passing, go on from there to make a good story. His utter craftsmanship, his training, and his personality are all forgotten. He is pictured as an old rummy who carved an eagle and swapped it in the Portsmouth, New Hampshire grog shops, across the river from dry-state Maine, for alcoholic beverage.
>
> Well, he did in his later years; but not a tenth as often as the general run of present day talk would have it. John Haley Bellamy drank to subdue a whirling brain.[28]

Regardless of how few or many bottles of rum he had consumed, by 1910 the seventy-four years John Haley Bellamy had spent upon this earth finally caught up to him. In the last known photograph to have been taken of the great woodcarver, believed to hail from that very year, a visibly haggard Bellamy looks into the camera [Fig. 9.3]. Like an old sea captain who has refused to give up the helm of his ship, he was still at work in his shop, defiant in the face of his years, proving to us again that, despite the words of his nephew, he had not abandoned his calling three years earlier on the death of Captain Lawrence. By the end of the year, things had taken a turn for the worse in Bellamy's life. On December 5, 1910 Doctor E. B. Eastman of Portsmouth, New Hampshire, was called in to examine Bellamy, and testified:

> This is to certify that I have this day examined Mr. John H. Bellamy and find him to be of unsound mind, incapable of managing his estate and protecting his rights..[29]

Deemed legally incompetent, another doctor, Dr. J. D. Carty, was appointed as Bellamy's legal guardian, and by February 1, 1911, John Haley Bellamy was removed from the Pepperrell Mansion. Due to "failing health and the need of personal care" he was compelled to spend his last days in the care of his cousin Harriot Mudge and her husband George in nearby Portsmouth.[30]

Of his time with the Mudges, we know precious little. Past writers have sought to depict life under their roof in pleasant terms, describing their home as "a small, comfortable three-story brick townhouse with two rooms on each floor," and assuring their readers "John's personal records and cherished possessions were moved with him."[31] They paint a stark portrait of Bellamy himself; "nothing remained of the once dashing, carefree, industrious artist. At seventy-five, he had deteriorated and was senile."[32]

About the last days of his life, questions and mysteries remain. Exactly how far had Bellamy's mind deteriorated? While he was growing senile by that point in his life, we also know that Bellamy was still capable of conceiving, planning, and executing a complex carved chest for his nephew in 1912, two years after he had been deemed in need of guardianship. (The feat also required that he travel back and forth to his old studio in Kittery, as that was where his tools were, and that is where the chest was discovered.) A smattering of receipts from the years leading up to Dr. Eastman's verdict inform us that despite his age Bellamy was still successfully engaging in the daily rituals of life. In June 1906, Bellamy had dental work done, several of his teeth being filled and bridged.[33] In 1907 he was still purchasing coal and groceries next door from Frisbee Brothers Meats and Groceries, as his family had done since the day he was born,[34] and milk from the H. F. Colby dairy.[35] And for twenty dollars he rented box number one at the Kittery Point post office, directly across the street from his home.[36]

While Dr. Eastman's declaration that Bellamy was "incapable of managing his estate" may be open to interpretation and debate, one thing that cannot be denied was his assertion that the aged woodcarver was powerless when "protecting his rights," for all was not well within the "small, comfortable" home of the Mudges. Bellamy's old friend George Savary Wasson noted that things were amiss in the Mudge household. In a letter sent to either Moses Victor Safford or (more likely) his father, the attorney Moses Safford, Wasson states:

> I am sorry to miss the meeting, but did not receive your message until after four o'clock, where under the existing circumstances it was too late.
>
> I hope you will call when down this way: I would like to talk over a matter concerning our friend J.H.B. with you.
>
> There is no doubt in the minds of certain of his friends here, including myself, that he is being cunningly and systematically bled by a party from Portsmouth, who apparently has no other such lucrative business to attend to.[37]

Wasson's suspicions were correct. *Schedule B of the First Account of John D. Carthy Guardian of John H. Bellamy of Kittery, in the County of York*, shows that the Mudges received a kingly sum for their services. Bellamy was paying anywhere from $49.50 to $80.10 a month for board and expenses, with additional payments for medical attention and winter clothing drawn separately.[38] In the decade from 1910 to 1920 the average household income in the United States amounted to only $750 a year.[39] In just two years of boarding Bellamy, Mrs. Mudge (Mr. Mudge having died by 1913[40]) accrued three full years' income—"$2,228.80—on top of her earnings as a dressmaker.

Operating from the false notion that Bellamy had been a drunkard, past authors have said that "because of his [Bellamy's] many years of dissipation, no great wealth was involved"[41] in regards to his estate. Such could not be further from reality. When Bellamy moved in with the Mudges, he had $3,362.67 in his possession, most of which resided in the annuity he had taken out, a measure that seems to have been designed by the artist as a self-generated pension of sorts. Yet in just two years, nearly two-thirds of his life savings were liquidated, with no obvious reasons being offered for why he had to pay out more money per month than most homeowners at that time paid for their mortgages. With his medical and clothing expenses a separate matter, the only other costs Bellamy's presence would have drawn were for food and basic necessities—two expenditures that in no way could have swallowed as much of Bellamy's savings as they did.

Locked in a mind and body that failed him more every day and being "cunningly and systematically" drained of his life's earnings, Bellamy yet found the strength and wit to pen one last poem. Scribbled down by Bellamy upon his night slate the very evening before his passing, it is claimed, the simple quatrain, thick with brooding and ominous imagery, illustrates the dire situation he found himself in.

> There's sharks a-lurking in the sea
> There's wolves within the wood
> There's buzzards perched on many a tree
> In a carnivorous brotherhood[42]

Fig. 10.1 Anonymous, *Eagle Plaque*: carved wood with original paint decoration, ca. 1915. [12 x 48 in.] *Private collection, AKA*. Crafted after Bellamy's death by an unknown Maine woodcarver, such pieces reveal his continuing influence on artisans in his home state. Note the eagle's banner features Masonic symbols, marrying two distinct periods of Bellamy's career.

Epilogue

FOR THE AGED, DECREPIT JOHN HALEY Bellamy, his long-lived prayer for release was answered in the seventy-eighth year of his life. When Atropos finally drew near to him on April 5, 1914,[1] his death did not go unnoticed in the Portsmouth, New Hampshire newspapers, garnering mention the next day in both the morning and evening papers:

> John H. Bellamy of Kittery Point, died at the home of Mrs. George A. Mudge on Sheafe street on Sunday evening after a lingering illness. He was born at Kittery Point April 16, 1836 and was the last of his family. He learned the carver's trade with Samuel Dockham [*sic*] of this city and soon became an expert at wood carving and his handiwork is to be found in many of the homes in Kittery and this city. He was of an inventive genius and in early life designed many patents from which others realized substantial revenues.
>
> He began his work as an expert woodcarver at the Portsmouth navy yard during the war of the rebellion and he was employed there until 1900. While at the navy yard he made the figure heads and did other carving for the war vessels constructed during that period. His services were highly valued by the late Naval Constructors Wilson and Hichborn. In addition to his work on the navy yard he carved the figure heads for many of the clipper ships built here when shipbuilding flourished on the Piscataqua river.[2]

and

> John H. Bellamy, a native of Kittery Point died on Sunday at the home of Mrs. George A. Mudge, Sheafe street, at the age of 77 years. He had been ill for several months and had resided with the Mudges for a few years who cared for him.
>
> Mr. Bellamy was a noted wood carver, learning it with the late Samuel M. Dockum, and specimens of his fine work have often been shown. He was also engaged in figure heads when these were in vogue. He was extremely skilful [*sic*] with his tools, and was the last of the family.[3]

According to Safford, the cause of death was of "general senile deterioration" (MVS: *Eagle*, Pg. 5). The State of New Hampshire recorded Bellamy died due to apoplexy, the archaic term for a stroke.[4] On Wednesday afternoon, April 8, at 2:30 in the afternoon, funeral services for the great woodcarver were held at Kittery Point's Free Will Baptist Church, the same church his father had so devoutly attended and served. It is not recorded how many (or few) attended the somber occasion that was officiated by R. W. Churchill. The service concluded with Mr. H. W. Nickerson overseeing the interment of Bellamy's remains in the cemetery immediately behind the church within the Bellamy family plot.[5] Local papers made certain to cover the event.

> John H. Bellamy, who was buried at Kittery Point today, was a man of genius. He received many patents from the government for inventions. He had literary and

> poetical ability, was strong in friendship, and always extended a welcoming hand to returning sons of Kittery Point. Mr. Bellamy had been ill for several years, but his death came as a shock to his friends. Mr. Bellamy was the son of the late Hon. Charles G. Bellamy, formerly state treasurer of Maine, and a constructor at the Portsmouth navy yard.[6]

With a handful of prayers and a few paragraphs in the local papers, the life of one of America's supreme woodcarvers come to a close. Yet as the years moved forward in Kittery, the memory of John Haley Bellamy persisted. Whether it was his "quizzical smile" or his fancy waistcoats, his penchant for mischief, his love for "appetizing cookery," or even his artwork, the memory of John Haley Bellamy would not easily vacate the minds and hearts of his neighbors. From time to time they would find their thoughts straying back to him, sighing wistfully as they recalled his extravagant ways and quirky habits, remembering him in purely Kittery terms as being "as independent as a hog on ice,"[7] words thick with respect for a man they never did quite comprehend. And though his tools now lay silent, the works he had forged with them continued to inspire local artists, moving them to craft eagle plaques akin to his work, even though the man who had popularized such designs was now no more [Fig. 10.1].

Work as accomplished and beautiful as Bellamy's could not be forgotten. Nor could it be ignored. They had captivated the eyes of so many admirers when he was alive, and Bellamy's eagles continued to speak for their creator while he slept in eternal repose. Like the proverbial phoenix rising from the ashes, so did Bellamy's artistic legacy experience its rebirth. It did not take long. While many an artist has had to linger in a malaise for nearly a century or more after their passing to be rediscovered, Bellamy's legacy had to wait a scant twenty years. It began with the Mariner's Museum's purchase of the *Lancaster* figurehead. With such a large and dramatic piece of sculpture now prominently displayed upon their premises, the Museum inevitably sparked a genuine interest in not only the *Lancaster* Eagle but also in the man who crafted it. A groundswell of curiosity among the general public spurred its new owners into researching its history and discovering its origins, an act that inevitably led them to the doorstep of Moses Victor Safford.

The letters and two biographical sketches Safford provided to the Mariner's Museum concerning his uncle were not only the first attempts at documenting the creation of the *Lancaster* Eagle but also served as the first effort to capture for posterity the rich life story of John Haley Bellamy. And as the public's interest in Bellamy continued to build, Safford found himself moved to author yet another piece on his uncle. "John Haley Bellamy, The Woodcarver of Kittery Point," appeared in the March 1935 issue of *The Magazine Antiques*. Accompanied by an editorial titled *Bellamy's Style and Its Imitators* by Homer Eaton Keyes, the magazine's editor, Safford reintroduced Bellamy to a national audience in earnest, ordaining his handiwork as meriting a collector's attention.

As with any great scholarly inquiry, Safford's article raised far more questions than its author could ever answer. Having relied almost exclusively upon personal recollections drawn from family and his own youth, and believing that Bellamy's letters, poems, photographs, and other priceless ephemera had been irrevocably lost, Safford found himself unable to furnish any further information, even as the public increasingly clamored for more. A fortuitous discovery would be made in the 1940s, when the Pepperrell Mansion passed into the hands of Kittery's Frost family. Distant descendants of the home's storied first owners, one member of this family, the inquisitive Joseph William Pepperrell Frost, would revolutionize our understanding of John Haley Bellamy when he entered Bellamy's old workshop in the second floor of the old Pepperrell counting house.

> It was a little over 60 years ago that I came home from the army to Kittery Point. I remember that the train brought me to Portsmouth and I waited for the Kittery Point bus because I hadn't been to Kittery Point in some years and my father lived at the Pepperrell Mansion. . .The next day I went down to Bellamy's workshop . . . in the old counting house by the river.

My father was going to rip that down; its destruction was already scheduled. I went into the place and downstairs it looked like just about any workshop that you'd run into. Upstairs, there were workbenches that John Bellamy used and they were still covered with things he had left on them. I remember, there was an old stuffed eagle over in one corner and more tools and pieces of woodwork left around. There were also endless papers and I decided that I was going to have that building for myself. I should have locked the door and left it, right then. But instead I began cleaning it up. [8]

As if in a scene lifted from *Raiders of the Lost Ark*, Frost found himself surrounded by an antiquarian's wildest dream. Bellamy's tools lay untouched, exactly where he had placed them the last time he had ever cut into a block of wood. Unfinished sculptures sat upon his bench; various completed masterpieces lay scattered about the room; the cartoons of Thomas Nast hung in yellow tatters from the walls. Stuffed into a trunk or two he found the lost personal correspondences of John Haley Bellamy and family. (More artifacts would await him when, soon after, he entered the attic of the Pepperrell Mansion.)

Over a period of years I found some interesting things in there. I was probably there six months before I went into the back room of the place. But there in the back room, hanging up on the beams, were rolls of paper. I took them out and found they were patterns of eagles that Bellamy had carved over the years. This was the workshop itself, and you could almost see him sitting at the bench there. The place in the floor beside the chimney still had sand from the time the stove sat there. The door going into the place had the marks of different friends of his, like Wasson and Howells, written on it, but I don't know why. I later painted it out. I'm not sure the door is even there now. [9]

Joseph Frost had stumbled upon a treasure trove of priceless information. And well he knew it. Yet rather than publish his findings outright, the humble historian chose to disseminate the information and artifacts he had unearthed, seeking to place them into the hands of those collectors or museums he felt best capable of preserving and propagating the memory of John Haley Bellamy.

Interest in Bellamy's handiwork had not abated since Safford's article in *Antiques*, and with the new wave of patriotism flowing throughout a triumphant post-Second World War America, outfits like the Artistic Carving Company, with their mass-produced copies of his eagles, fueled a fascination in all things Bellamy. Scholars like Marion Brewington, Erwin Christensen, and Pauline Pinckney included biographical blurbs about Bellamy in their histories of American marine woodcarving. Kittery resident Horace Mitchell, authored articles about Bellamy's life in such publications as the *American Mercury* and *Yankee Magazine*, perhaps at the suggestion of his neighbor down the road, Joseph Frost.

As the arrival of the American Bicentennial grew imminent and an interest in all things Americana inevitably arose, Phillip Isaacson published a book detailing the artistic evolution of that most American of symbols, the eagle. Sensing the grand opportunity this presented, Frost went to work again, quietly advocating for Kittery Point's favorite son. With a generous assistance that included making his personal collection of Bellamy sculptures and paper patterns available to Isaacson, Frost's lobbying efforts gained Bellamy an entire chapter in the historian's book, an inspired piece that was destined to become an instant classic on the subject of the eagle in American art. Frost had scored a decisive coup. The chapter was replete with photographs of Bellamy and his works, written in Isaacson's authoritative, competent hand. Frost eagerly sought to capitalize on it, again magnanimously sharing his expertise and collection to further the public's awareness of Bellamy via two high-profile exhibitions at the William A. Farnsworth Library and Art Museum of Rockland, Maine, and the Maine State Museum. Titled *The Heritage of Woodcarving* and *Versatility—Yankee Style*, these shows, rich with the many varying examples of his craft and superbly curated, were true landmarks, especially since only one other public exhibition of Bellamy's diverse creations had ever been arranged, back in the 1960s.[10] Bellamy was finally gaining his just due.

Throughout the 1970s and into the 1980s, museums and private individuals well beyond the precincts of New England began to actively collect examples of Bellamy's handiwork in ever-increasing numbers. Moses Victor Safford and Homer Eaton Keyes may have made it permissible to collect Bellamy, but Joseph Frost made it desirable. Exhibitions were held and articles written about Bellamy. An entire chapter in a widely circulated book had even been devoted to him Even so, until the 1980s no one had yet attempted to pen a definitive biography about Kittery Point's favorite son. In 1982 journalist Yvonne Brault Smith wrote *John Haley Bellamy, Carver of Eagles*, the first monograph for Joe Sawtelle's new Portsmouth Marine Society.[11] Hers was the first attempt at capturing Bellamy's life and career in detail, stretching beyond a mere survey or commentary upon the man and his sculptures. Smith was the first to make any use of those Bellamy family correspondences that Frost had discovered four decades earlier.

Today, due to the efforts of many historians, authors, collectors, and curators, John Haley Bellamy is a name recognized far beyond his native Kittery. For those who wish to view his handiwork, several of America's finest museum's grandly accommodate the curious. To many maritime buffs and Bellamy aficionados, a trip to the Mariner's Museum in Newport News, Virginia, to see the *Lancaster* Eagle figurehead takes on all the character of a pilgrimage. For those who brave the delights of a long drive along Maine's spectacular coastline, a visit to the William A. Farnsworth Library and Art Museum in Rockland, home to the largest and most pertinent collection of Bellamy artwork, artifacts, letters, and ephemera, provides a rich reward. Any number of the great woodcarver's masterpieces can be found in the galleries, rubbing shoulders with the works of such artists as Fitz Henry Lane, Thomas Cole, the Wyeth family, and even Bellamy's old friend Winslow Homer. Another Maine institution promoting an appreciation of Bellamy's abilities is the Kittery

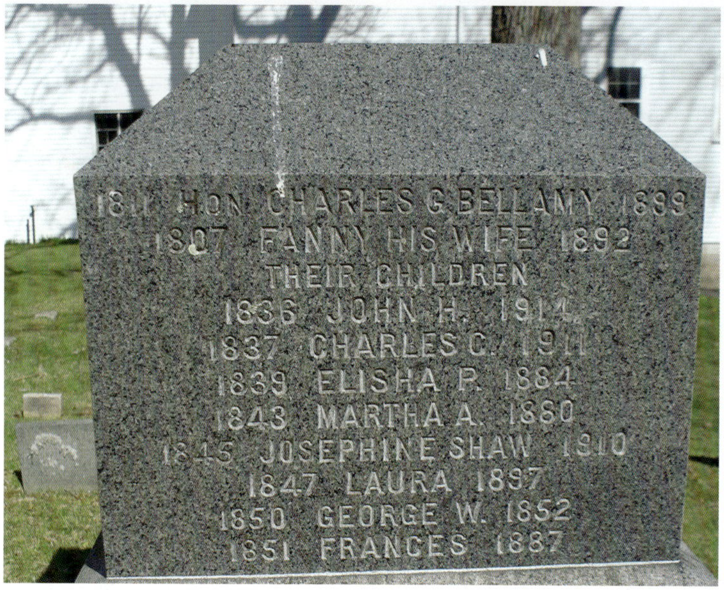

Fig.10.2 The Bellamy family tombstone with Kittery Point Free Will Baptist Church in the immediate background. *KC, 2008.*

Historical and Naval Society's museum. Though small in size and reputation, this charming cultural gem of the New England coast enjoys a surprisingly deep collection. Under the leadership of curator Wayne Manson, the collection tells the story of Bellamy and his hometown to tourists from around the world. The Museum of Our National Heritage in Lexington, Massachusetts, serves as the prime repository for Bellamy's Masonic and Odd Fellow clock cases and frames, offering patrons the chance to treasure the most underappreciated phase of his career. And to gain an intimate understanding of just how he wrought so many splendid sculptures from blocks of black walnut, mahogany, and pine, a trip to the Peabody Essex Museum in Salem, Massachusetts, offers the chance to view Bellamy's tool chest alongside some of his finest creations.

Bellamy's influence is still quite palpable in his old stomping grounds. Indeed, if he were to find himself walking the streets and shady lanes of his native Kittery Point and nearby

Fig. 10.3 The Ann Treadwell House, now Northeast Auctions, Portsmouth, N.H, with a massive reproduction Bellamy eagle by carver Ron Raiselis akin to those once on Portsmouth businesses. *CO, 2013.*

Portsmouth, he might see neighborhoods that have changed little over the past century. The venerable Pepperrell Mansion and its counting house/studio still stand today, preserved by its owner in a tasteful fashion. The wharves beyond Bellamy's shop windows still hum with activity, though sail-driven brigs and schooners have long since been replaced by diesel-powered lobster boats. Every Sunday the house of worship that the Bellamy family attended beckons the faithful to services, just as it did so many years before, while in the very shadow of the church the family's final place of repose can be found [Fig. 10.2]. And just down the street from where he lived, the Portsmouth Navy Yard is still in full operation today, proudly servicing, upgrading, and repairing the warships of America's submarine fleet.

Bellamy eagles still adorn local businesses [Fig. 10.3] and homes [Fig. 10.4], though now reproductions suffice in place of the master's handiwork. And in what is surely the most fitting tribute to his genius, the art of John Haley Bellamy still inspires woodcarvers today. While businesses churning out "Bellamy-esque" reproductions can be found throughout New England, individual artists devoted to capturing the nuances of Bellamy's art in all its discipline and glory are found in Kittery and its immediate environs. Indeed, it seems the closer one draws to Kittery, the more devoted to Bellamy these artisans become.

Immediately down the road from Kittery, in the neighboring town of York, Maine, can be found the studio of Artie Swanson [Fig. 10.5]. Swanson's Cider Hill Woodworks, Inc. has earned the reputation of being the undisputed first choice among collectors wishing to own museum-quality copies of Bellamy sculptures, from Bellamy's magnificent outdoor eagles to his indoor plaques and traditional banner-toting raptors [Figs. 10.6-10.7]. Originally a cabinetmaker, sign carver, and repairer of antique furnishings by trade, Swanson became interested in Bellamy's handiwork back in the 1970s while toiling within a

Fig. 10.4 (top) Reproduction Bellamy eagle by Arthur Swanson adorning a Kittery-area home. *AS.*

Fig. 10.5 (right) Cider Hill Woodworks, Inc., and its master woodcarver, Arthur Swanson. *CO, 2013.*

cooperative woodshop in Portsmouth, New Hampshire. Swanson recalls his fateful first encounter with a Bellamy eagle:

> A client brought a 24" white eagle in for repair of the tips of the wings and beak. This was at the time when Bellamy eagles were starting to be rediscovered. I was taken by the simplicity of lines and the gracefulness of the bird. After repairing several, I tried my hand at carving some... and over time I have come to respect Mr. Bellamy's eagles as some of the most beautiful examples of American folk carvings.[12]

By 1981 Swanson had built a small shop in York, Maine; he has found himself carving a steady stream of signs, plaques for schools, and Bellamy reproductions ever since. Through years of studying and painstakingly measuring Bellamy's actual handiwork and placing scientific precision into his copies of Bellamy eagles, Swanson has come to produce pieces that are virtually indistinguishable from the originals, undoubtedly much to the future consternation of museum curators and antiques dealers [Fig. 10.8].

Another Bellamy devotee can be found a stone's throw down the street from Swanson's workshop. While Mr. Swanson may produce the most authentic reproduction Bellamy carvings, remaining unerringly true to the deceased artist's imaginative vision, architectural woodcarver Michael Dow is known as the

Fig. 10.6 (top) Arthur Swanson, *Reproduction of a John Haley Bellamy Eagle Plaque*: painted and gilded pine, 2009. [20¾ x 43½ x 3 in.] *DB*.

Fig.10.7 (middle) Arthur Swanson, *"Fortis Sed Non Ferox"* eagle: painted pine, 1991. [8¾ x 26 x 3 in.] *DB*.

Fig. 10.8 (bottom) Two examples of the Bellamy family coat of arms, one by John Haley Bellamy (right), the other by Arthur Swanson (left.) Once the reproduction on the left is stained, the two will be virtually indistinguishable. *AS*.

purveyor of some of the most original creations crafted in the spirit of John Haley Bellamy's handiwork [Fig. 10.9]. Declaring that every woodcarver in the Kittery area is "doomed to get caught up in Bellamy"[13] at some point in their career, Dow engages in reproductions and repairs of Bellamy eagles, and he has also seen fit to take Bellamy's artwork in directions never before imagined. Rising to the challenge of creating an eagle true to Bellamy's aesthetic principles yet durable enough to withstand the elements almost indefinitely, Dow struck upon the novel solution of having Bellamy eagles sandcast out of sturdy aluminum and coating them in marine paint. The final product is a sculpture perfect for outdoor display [Fig. 10.10]. Dow has also turned his hand to creating men's jewelry fashioned in the designs of Kittery's favorite son. The road that led him to this unusual take on the Bellamy eagle began, oddly enough, when Mr. Dow was rear-ended by a quadriplegic motorist behind the wheel of a van. As Dow recalls:

I was doing rather heavy woodworking at the time of the accident. As a result of it, I could not lift the heavy pieces any more because the strain of it all made my arm worse. That is when I started to think small. I had been

Fig. 10.9 (top) Michael Dow, expert architectural woodcarver, creator of handmade bagpipes and crafter of inventive Bellamy-themed creations. *CO, 2013.*

Fig. 10.10 (bottom) Sandcast aluminum reproduction of a Bellamy eagle, manufactured by Michael Dow. *MD / CO.*

> doing regular size wooden Bellamy eagles for quite a while and I kept making them smaller. Finally, I carved, gilt and painted one about as small as the jewelry and thought: "There has just got to be a better way to make these." The jewelry thing seemed like the logical next step as they were just that size.[14]

Issued as a line of pins and tie tacks, the final product of Dow's handiwork is replete with exquisite, charmingly expressive detail [Fig. 10.11].

Michael Dow proves how adaptable Bellamy's eagles are to modern times and that Bellamy's vision is alive and well today. In a tradition hailing back to the Spanish-American War, when Bellamy eagles sported the patriotic slogans "Remember the Maine" and "Long Live Dewey," so did Dow elect in early 1991 to craft a Bellamy eagle bearing the motto "Free Kuwait!" Intended as a gift for President George Bush Sr. and presented as an expression of Dow's support for Operation Desert Storm via Maine senator William Cohen, [Fig. 10.12] the piece would be featured in the 1996 National Archives exhibition *Tokens & Treasures: Gifts to Twelve Presidents*, one of only twelve gifts selected from the thirty-thousand-plus presents that were given to President Bush during his presidency.[15] (Dow repeated this gesture in the days immediately following September 11, 2001, crafting and presenting a Bellamy eagle with the saying "United We Stand!" to President George Bush Jr.)

Like so many of Kittery Point's seafarers from generations past, lobsterman and woodcarver David Kaselauskas, when not busy harvesting King Neptune's riches turns his talents to producing Bellamy-inspired artifacts for modern use. His hand-carved wooden or bronze-cast cane handles are meticulously crafted to resemble the head and neck assembly of a classic Bellamy eagle. His limited edition canes [Fig. 10.13] and walking sticks are found in the hands of a select number of collectors. And in a fashion little different from Bellamy himself, this product of coastal Maine's maritime artistic heritage has turned his woodcarver's tools beyond eagles, crafting reproductions of Bellamy and Wasson's elaborately carved wooden boards. Largely self-taught, Kaselauskas was particularly drawn to Wasson's rich artistic legacy by his personal friendship with Joseph W. P. Frost, [Fig. 10.14] who,

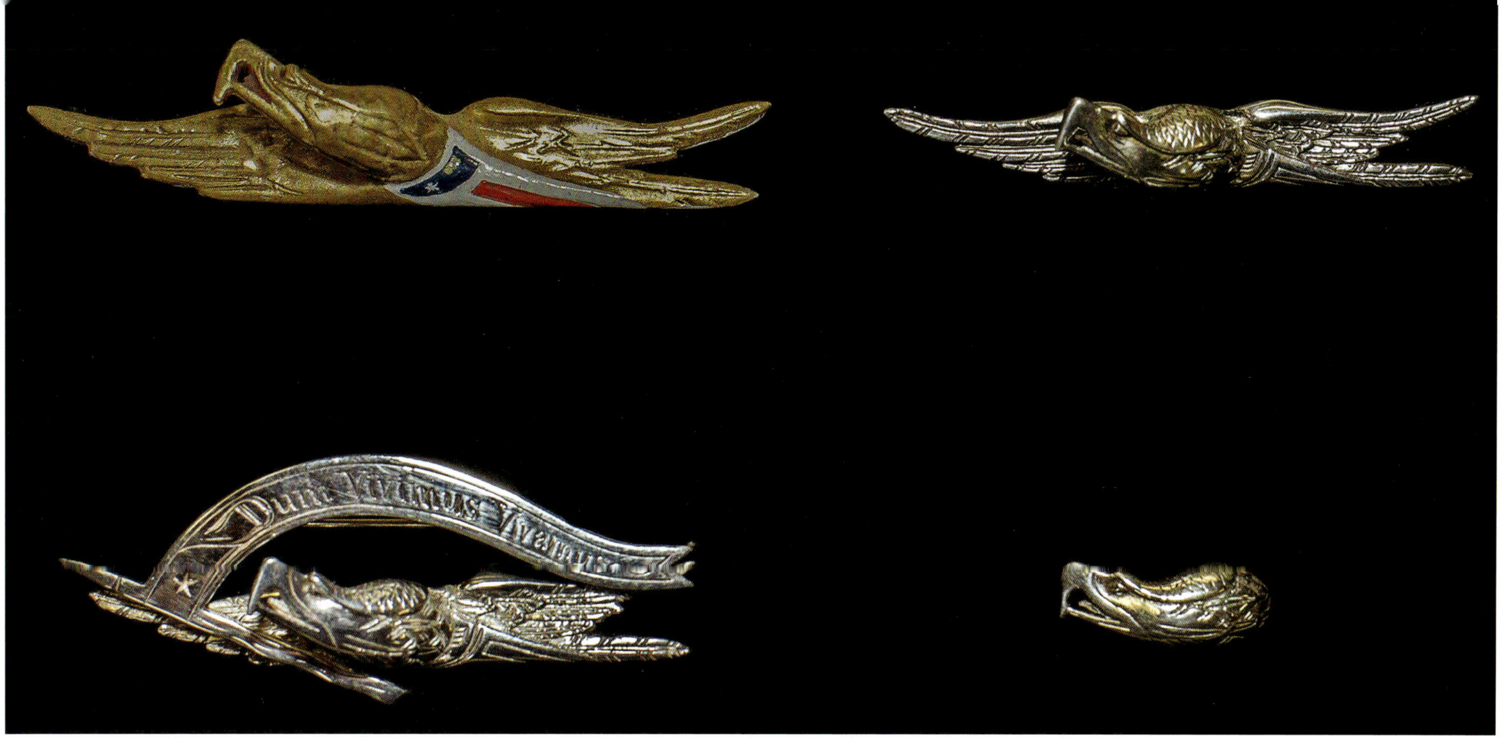

Fig. 10.11 Michael Dow, *Bellamy Eagle*: painted pine, 1990 [3 in. long], most likely the smallest rendering of the Bellamy eagle ever achieved, and examples of Dow's Bellamy Eagle-inspired lapel pins, tie pins, and tie tacks. *MD / CO.*

as he had done for Bellamy, sought in his last years to generate interest in Wasson. Kaselauskas' wooden boards have drawn the notice of several of today's noted marine artists, including Robert Grimson, Christopher Ward, John Morrison, and Cheri Dennett, who all use them as a fitting platform for hosting their paintings [Figs. 10.15 and 10.16].

Lest we think of Bellamy's modern devotees as being limited solely to Kittery and its surrounding environs, British-born artist Vinnie Harrild has crafted novel and fitting twists to the traditional Bellamy eagle. First introduced to the work of John Haley Bellamy back in the 1970s when he was a restorer of antiques and fine art, Harrild found that some of the Bellamy woodcarvings he was hired to restore "were in such a dreadful state of disrepair" that re-carving was required.[16] From that experience, Harrild was compelled to learn Bellamy's preferred carving technique and adopt his approach, a happy turn of events that has placed him in the enviable position as today's premier choice among private collectors for the restoration of damaged Bellamy carvings.

He declares, "To enter into a carving project is to enter into a contract with the original artist. It's impossible to work on an eagle without some kind of dialogue with Bellamy." Harrild's mastery of Bellamy's techniques has led him into a unique partnership with America's finest clockmaker. Harrild and Chelsea Clock have done nothing less than combine two distinct phases of Bellamy's woodcarving career into a unique work of art, intended as an homage to Bellamy rather than as a faithful rendering of his handiwork [Fig. 10.17]. Taking the vibrant design of Bellamy's *Eagle plaque* and transforming it into a handmade clock case in which to house one of Chelsea Clock's world-famous timepieces, the final product is a triumph, embodying a hint of the great woodcarver's whimsical nature and retaining "the simplicity of Bellamy's statement on patriotism," while giving us an excellent idea of what might have happened if *Titcomb & Bellamy* had ever strayed beyond Masonic-themed work.[17]

Fig. 10.12 (top left) Michael Dow's *Free Kuwait!* Bellamy Eagle, as it appeared on display at the 1996 National Archives exhibition *Tokens & Treasures: Gifts to Twelve Presidents*. *MD.*

Fig. 10.13 (left) The Kittery Point workshop of lobsterman/woodcarver David Kaselauskas, complete with several examples of his Bellamy-inspired walking sticks. *DB, 2013.*

Fig. 10.14 (top right) David Kaselauskas (left) with Joseph William Pepperrell Frost (right), rediscover of John Haley Bellamy, in Frost's study: photograph, 2006. *DK.*

Nearly a century after his passing, John Haley Bellamy continues to inspire woodworkers and the general public alike. His is a tradition that, unlike the offerings of so many other artists from bygone years, has defied the changing temperaments and fashions of the fickle public. Instead, Bellamy's is a *living* tradition, one that continues to breathe under the appreciative eyes of discerning fine art collectors and the exertions of woodcarvers who every day turn out sculptures directly inspired by his artistic vision. These creations grow with and adapt to the needs of today's Americans in a fashion no different from the way Bellamy's handiworks served the needs of Americans back in his time. Whether beneath the massive wings of his *Lancaster*

Fig. 10.15 John Morrison, *"Loella" in Chauncey Creek*: oil on carved walnut board, 2008. [10 x 13½ in.] *DB*.

Fig. 10.16 Cheri Dennett, *Wood Island in October*: oil on carved pine board, 2008. [11 3/16 x 14 5/8 in.] *DB*.

figurehead or in the proud visages of his classic Bellamy eagles—even within the deft expressiveness of his more atypical carvings—the "overwhelming *joie de vivre*" that was Bellamy's America can be glimpsed by today's Americans. To gaze upon one of Bellamy's raptors is to behold our nation when it was filled with a common purpose, triumphant in its accomplishments, confident in its abilities, and convinced of its destiny as the leader of nations. To look into the fierce eye of a Bellamy eagle is to look into the very soul of Bellamy's America—a soul that, like his art, is still very much alive and active today.

Fig. 10.17 Vincent Harrild and Chelsea Clock Co. *Chelsea Clock Bellamy Eagle*: antique Eastern white pine and mahogany (both approximately two hundred years in age), 23-karat gold leaf, 2011. [17 x 19 x 8 in.] *Courtesy of Vinnie Harrild and Chelsea Clock.*

Endnotes

Introduction

1. Typewritten copy of personal correspondence from John Haley Bellamy to Charles Gerrish Bellamy, March 3, 1866 (Acc# 1949.217), collection of the Farnsworth Art Museum and Wyeth Center
2. Horace Mitchell: "Eagles with Feelings." *Yankee Magazine*, September, 1965, Pg. 28
3. http://www.pbs.org/wgbh/roadshow/fts/providence_bellamy eagles.html
4. Moses Victor Safford, *John Haley Bellamy, The Woodcarver.* Unpublished manuscript, November 23, 1941, collection of the Mariner's Museum, Newport News, VA., Pg. 3 (hereafter cited as MVS: *JHB*)
5. *Explanation of Plan of Old Pepperrell House, Kittery Pt. ME.* Unpublished, undated manuscript, collection of Dyer Library, Saco, Maine
6. A fourth institution, the Rice Public Library of Kittery, Maine, while having no actual Bellamy family letters in its collection, should also be mentioned; it holds several Xerox copies of letters sent to John Haley Bellamy and Charles Gerrish Bellamy that presently reside within private collections.

Lancaster 1

1. Extracts from the *Daily Log Book,* US Navy Yard, Portsmouth, New Hampshire, October 15, 1819 – December 17, 1929. Compiled by Captain David F. Boyd, USN., United States Naval Prison, Portsmouth, NH, March 31, 1931, July 12, 1875
2. http://www.history.navy.mil/danfs/l3/lancaster-i.htm
3. Richard E. Winslow III, *"Do Your Job!" An Illustrated Bicentennial History of the Portsmouth Naval Shipyard, 1800–2000.*The Portsmouth Marine Society (published for the Society by Peter E. Randall), Portsmouth, NH, 2000, Pgs. 41–42
4. Extracts from the *Daily Log Book*, July 12 and 22, 1875; *Charlestown Navy Yard: Boston National Historical Park, Massachusetts.* Produced by the Division of Publications, National Park Service, US Department of the Interior, Washington, DC, 1995, Pg. 17
5. RG 181 Naval Districts and Shore Establishments, Portsmouth Navy Yard, NH: *Letters Sent to the Bureau of Construction and Repair, 1842-1911, Vol. 7, from Aug 11, 1875 to Dec 1, 1877.* Correspondence #180, from Earl English, Captain Commandant, Portsmouth Naval Shipyard, to Isaiah Hanscom, Chief Constructor, United States Navy, December 13, 1876, collection of National Archives and Records Administration, Waltham, MA

Chapter 1

1. "List of Bellamy Family Birth and Death Dates," Pg. 5 (Acc# 1949.185), collection of the Farnsworth Art Museum and Wyeth Center; MVS: *JHB*, Pg. 1; Moses Victor Safford, "John Haley Bellamy, The Woodcarver of Kittery Point." *The Magazine, Antiques*, Vol. XXVII, No. 3, March, 1935, Pg. 102 (hereafter cited as Safford, Moses Victor, *The Magazine, Antiques)*; *Portsmouth Herald*, April 6, 1914. It should be noted that these documents correct the unsubstantiated claim that John Haley Bellamy was born on April 5, 1836. [See: Yvonne Brault Smith, *John Haley Bellamy, Carver of Eagles*. Portsmouth Marine Society, Portsmouth, NH, 1982 (reprinted 2002), Pg. 1]
2. Everett S. Stackpole, *Old Kittery and Her Families*. Pg. 18
3. John J. Babson, *History of the Town of Gloucester, Cape Ann*. Procter Brothers, Gloucester, MA, 1860 (Re-issued by Peter Smith Publisher, Inc. 1972), Pg. 21
4. Special thanks to David Kaselaukas of Kittery Point, ME, for his assistance in informing and explaining the many day-to-day terms and phrases that were used in Bellamy's time, which remarkably are still heard in conversation with some Kittery Point residents today. The meanings of these particular words and phrases are as follows: *krawm* and *sculch*—garbage; *plauggy-hubbly* and *chowy-hubbly*—two terms used to describe different forms of rough seas. *Black dreapin' fog*: a heavy gray fog, exceptionally thick and moist; *nail sick*—when a vessel's nails had begun to rust, causing its hull to become loose and leak profusely. "He's slipped his mooring": a phrase meant to inform someone that a person has died' another common way of declaring this—"The Maker has called him aft," "thick as mashed mud" is self-explanatory.
5. MVS: *JHB*, Pgs. 10–11. In this same passage, Moses Victor Safford claims that one of Bellamy's schoolmates became the colorful

adventurer "Buckman Pasha" of Turkish naval fame. As "Buckan Pasha" was born Ransford D. Buckman in Hantsport, Nova Scotia, in 1869 [*The New York Times*, May 30, 1915], it is impossible for him to have been schoolmates with John Haley Bellamy. It should also be noted that in her book *John Haley Bellamy, Carver of Eagles,* Yvonne Brault Smith makes the claim that the sword in question hung within the home of the Seward/Seaward family, attributing this information to Moses Victor Safford's writings at the Mariner's Museum, Newport News, VA. [See Yvonne Brault Smith, *John Haley Bellamy, Carver of Eagles*. Portsmouth Marine Society, Portsmouth, NH, 1982 (reprinted 2002), Pg. 45] Smith's claim is unfounded and unsubstantiated, as Safford never identifies the name of the family who owned the sword in question in his writings.

6. Mark Fernald, *Life of Elder Mark Fernald, Written by Himself*. George Moore Payne and D. P. Pike, Newburyport and Philadelphia, 1852, Pg. 14, as quoted within *The Heritage of Woodcarving: An Exhibition of Decorative Arts by John Haley Bellamy, Master Ship Carver and Seven Rockland, Maine, Carvers, July 18 through September 7, 1975*. The William A. Farnsworth Library and Art Museum, 1975, Rockland, Maine, Pg. 5
7. Ascribing the date of around 1700 for the construction of the Sir William Pepperell Mansion runs contrary to the long-held belief that this house was built in the year 1682. The dating of this structure to around the year 1700 is found in Richard M.Candee, "The Architecture of Maine's Settlement, Vernacular Architecture to About 1720" in *Maine Forms of American Architecture* (Deborah Thompson, ed.) pp. 28, 43-44. Dr. Candee found the Pepperrell Mansion revealed that, aside from a lone seventeenth-century beam in the cellar holding up the first floor (reused from a previous structure), no evidence of woodwork prior to the dawn of the eighteenth century is present. Candee hypothesizes that the home's interior paneling and stairs were not installed until the 1730s. [E-mail communication with Dr. Richard Candee, Prof. Emeritus, Boston University, November 18, 2012.]
8. Everett S. Stackpole, *Old Kittery and Her Families*, Pg. 246; and John D. Bardwell, *Images of America: Old Kittery*. Arcadia Publishing, an imprint of the Chalford Publishing Corporation, Dover, NH, 1995, Pg. 77
9. "Tourists, Attention" handbill advertisement, n.d. (Acc# 1949.113), collection of the Farnsworth Art Museum and Wyeth Center; Everett S. Stackpole, *Old Kittery and Her Families*, Pg. 246
10. Pauline A. Pinckney, *American Figureheads and Their Carvers*. W.W. Norton & Company, Inc., New York, 1940, Pgs. 46–47
11. Moses Victor Safford, *Bellamy Genealogy*. Unpublished manuscript dated January 29, 1938, collection of Dyer Library, Saco, ME
12. Moses Victor Safford, *Bellamy Genealogy*.
13. Moses Victor Safford," *The Magazine, Antiques*, Pg. 102
14. *Charles G. Bellamy*. Unpublished, undated manuscript, collection of Dyer Library, Saco, ME
15. Unpublished, un autobiographical notes penned by Charles Gerrish Bellamy, collection of Dyer Library, Saco, ME
16. Moses Victor Safford, *Bellamy Genealogy*. Charles Gerrish Bellamy appears to have only held this posting for one year, 1842.
17. *Charles G. Bellamy*.
18. *Charles G. Bellamy*.
19. *Charles G. Bellamy*.
20. Unpublished, undated autobiographical notes penned by Charles Gerrish Bellamy, collection of Dyer Library, Saco, ME. Moses Victor Safford takes exception to this claim, stating that Charles Gerrish Bellamy's "characterization of himself as a 'house and ship carpenter' was evidently for political vote getting purposes," adding "I doubt if he ever worked a day in his life as a ship carpenter." [Source: Personal correspondence from Moses Victor Safford to Frederick Hill, June 29, 1946, collection of the Mariner's Museum, Newport News, VA.] Despite Safford's take on Charles Gerrish Bellamy's words, the volume of correspondences addressed to and from Charles Gerrish Bellamy involving requisitions for supplies and concerning both the politics of the Portsmouth and Charlestown Navy Yards and his employment within these yards that are held within the collections of the Farnsworth Museum and the Dyer Library contradict Safford's assumptions and prove that Charles Gerrish Bellamy did indeed work as a boat builder and ship carpenter.
21. The unpublished, undated autobiographical notes penned by Charles Gerrish Bellamy, collection of Dyer Library, Saco, ME spells his teacher's name "Curtze" but Charles Bellamy's own inscription in his copy of Edward Shaw's *Civil Architecture* (2nd ed. Boston: Marsh, Capon & Lyon, 1834) calls him "Professor John Kurtze the celebrated Architect and a Dane." This is consistent with recent findings by Earle G. Shettleworth Jr., Director of the Maine Historic Preservation Commission and Maine State Historian. who identities Charles Gerrish Bellamy's teacher as John Kutts, an architect originally hailing from Denmark who was listed in the *Boston Directory* from 1828 to 1838, and who continued to practice his trade in the towns bordering Boston until moving to Philadelphia in 1842. [Source: email exchange between Dr. Richard Candee, Earle G. Shettleworth Jr., and the author, October 23, 2013; Shettleworth, Earle G., Jr.: *Notes on John Kutts* Unpublished, undated manuscript, collection of Earle G. Shuttleworth Jr.]

22. *Charles G. Bellamy*; Moses Victor Safford, *Bellamy Genealogy*.
23. "Kittery had only two public schools until 1803, when the community was divided into school "districts of which the Island and Harbor formed the sixth." "Martin Luther Frisbee letter to his Alumni Association, *Island and Harbor Echo* Nov. 27, 1895 reunion issue, pg. 14.
24. Interview with Steven Estes, Kittery Point local historian, July 10, 2008
25. Moses Victor Safford, *The Magazine, Antiques*, Pg. 102
26. Moses Victor Safford, *The Magazine, Antiques*, Pg. 102
27. Moses Victor Safford, *The Magazine, Antiques*, Pg. 102
28. Moses Victor Safford, *The Magazine, Antiques*, Pg. 102. In the version of the tale where the object of Bellamy's affections actually married the lawyer in the story, it is stated by Kittery Point local Horace Mitchell that: "Love for a girl, a Kittery girl, entered John's life. He wrote poems to her – but she married a lawyer." [Source: Horace Mitchell, "Eagles with Feelings," Pg. 33] Given the abrupt, non-specific nature of Mitchell's claim, as well as the absence of any direct personal relationship between Horace Mitchell and the Bellamy family (unlike Safford, who obviously did enjoy such a relationship) and his failure to cite the source for his claim, the author has decided that Moses Victor Safford's rendition of Bellamy's failed courtship is the authoritative source, with Mitchell's to be regarded as merely anecdotal and in this instance flawed.

Lancaster 2

1. MVS: *JHB*, Pg. 13. It must be noted that in 1938 Safford stated that Bellamy was "left to take the initiative in finding opportunities in the Navy for his art." [See Moses Victor Safford, *The Eagle, Figurehead of the U.S.S. Lancaster*. Unpublished manuscript dated July 18, 1938, collection of the Mariner's Museum, Newport News, VA, Pg. 10, hereafter cited as MVS: *Eagle*.] It is the author's opinion that, given what we know today of how US navy yards operated in regard to contracting decorative woodworkers in the nineteenth century, combined with the fact that Safford's testimony in 1941 conforms to this knowledge, Safford's contradictory claims hailing from 1938 are often inaccurate and in general do not properly reflect Bellamy's employment at US Navy Yards.
2. RG 181 Naval Districts and Shore Establishments, Portsmouth Navy Yard, NHNH: *Letters Sent to the Bureau of Construction and Repair, 1842–1911, Vol. 7, from Aug 11, 1875 to Dec 1st, 1877*. Correspondence #180, from Earl English, Captain Commandant, Portsmouth Naval Shipyard to Isaiah Hanscom, Chief Constructor, United States Navy, December 13, 1876, collection of National Archives and Records Administration, Waltham, MA
3. The case for T. D. Wilson having specifically chosen Bellamy to make the USS *Lancaster* figurehead lies within the following statement, drawn from one of Bellamy's obituaries: "His [Bellamy's] services were highly valued by the late Naval Constructors Wilson and Hitchborn [*sic*]." *Portsmouth Herald*, April 6, 1914.
4. Pauline A. Pinckney, *American Figureheads and Their Carvers*. W.W. Norton & Company, Inc., New York, 1940, Pg. 30
5. Pauline A. Pinckney, *American Figureheads and Their Carvers*, Pg. 30

Chapter 2

1. Yvonne Brault Smith, *John Haley Bellamy, Carver of Eagles*. Pg. 7
2. M. V. Brewington, *Shipcarvers of North America*. Barre Publishing Company, Barre, Massachusetts, 1962, Pg. 88
3. Philip M. Isaacson, *The American Eagle*. New York Graphic Society, Boston, Massachusetts, 1975, Pg. 115
4. Pauline A. Pinckney, *American Figureheads and Their Carvers*, Pg. 31
5. Jane L. Port, *Boston's nineteenth century ship carvers*. *The Magazine, Antiques*, Vol. CLVIII, No. 5, November, 2000, Pg. 754
6. Trevor J. Fairbrother, *Painting in Boston, 1870–1930*, (essay) found within Trevor J. Fairbrother, *The Bostonians: Painters of an Elegant Age, 1870–1930*. Museum of Fine Arts, Boston/Northeastern University Press, Boston, Massachusetts, 1986, Pg. 31
7. Trevor J. Fairbrother, *Painting in Boston,* Pg. 43
8. *The Art Students' League of New York, Announcements for the Season of 1901–1902*. Collection of The Art Students' League of New York.
9. http://www.nationalacademy.org
10. Pauline A. Pinckney, *American Figureheads and Their Carvers*, Pg. 99
11. Jane L. Port, *Boston's nineteenth century ship carvers*, Pg. 756; Philip M. Isaacson, *The American Eagle*, Pg. 96
12. Jane L. Port, *Boston's nineteenth century ship carvers*, Pg. 756
13. Pauline A. Pinckney, *American Figureheads and Their Carvers*, Pg. 99
14. Philip M. Isaacson, *The American Eagle*, Pg. 96, 98
15. Jane L. Port, *Boston's nineteenth century ship carvers*, Pg. 757; M. V. Brewington, *Shipcarvers of North America*, Pg. 45
16. *Portsmouth Herald*, July 12, 1978

17. Personal correspondence from Charles Gerrish Bellamy to John Haley Bellamy, March 25, 1857 (Acc# 1949.244), collection of the Farnsworth Art Museum and Wyeth Center
18. Erwin O Christensen, *Early American Wood Carving*. The World Publishing Company, Cleveland, New York, 1952, Pg. 34
19. Pauline A. Pinckney, *American Figureheads and Their Carvers*, Pg. 31
20. *Bellamy Banter: The Life and Work of John Haley Bellamy*. Piscataqua Decorative Arts Society, Vol. III, 2007–2009 Lecture Series, 2011, Pg. 3
21. Erwin O Christensen, *Early American Wood Carving*, Pg. 96
22. Source: Interview with Daniel Finamore, curator of Marine Art, Peabody Essex Museum, June 4, 2008.
23. *Portsmouth Times*, April 6, 1914
24. *Portsmouth Herald*, April 6, 1914
25. R.W. Gerald D. Foss, *Samuel M. Dockum, Cabinetmaker*. Undated manuscript published by St. John's Lodge No. 1, F. and A.M., Portsmouth, New Hampshire, collection of the Portsmouth Athenaeum; Johanna McBrien, *The Furniture Craft Community of Portsmouth, New Hampshire, 1798–1837*. Thesis, Spring, 1996, collection of the Portsmouth Athenaeum, Pg. 122
26. R.W. Gerald D. Foss, *Samuel M. Dockum, Cabinetmaker*; Johanna McBrien, *The Furniture Craft Community of Portsmouth, New Hampshire, 1798–1837*, Pg. 149
27. Johanna McBrien, *The Furniture Craft Community of Portsmouth, New Hampshire, 1798–1837*, Pg. 144
28. R.W. Gerald D. Foss, *Samuel M. Dockum, Cabinetmaker*; Johanna McBrien, *The Furniture Craft Community of Portsmouth, New Hampshire, 1798–1837*, Pg. 149
29. Johanna McBrien, *The Furniture Craft Community of Portsmouth, New Hampshire, 1798–1837*, Pgs. 149, 151
30. Johanna McBrien, *The Furniture Craft Community of Portsmouth, New Hampshire, 1798–1837*, Pg. 151; R.W. Gerald D. Foss, *Samuel M. Dockum, Cabinetmaker*.
31. M. V. Brewington, *Shipcarvers of North America*, Pg. 143; Johanna McBrien, *The Furniture Craft Community of Portsmouth, New Hampshire, 1798–1837*, Pgs. 149, 151
32. *Portsmouth Journal*, February 28, 1852
33. Johanna McBrien, *The Furniture Craft Community of Portsmouth, New Hampshire, 1798–1837*, Pg. 147
34. Johanna McBrien, *The Furniture Craft Community of Portsmouth, New Hampshire, 1798–1837*, Pg. 140
35. Johanna McBrien, *The Furniture Craft Community of Portsmouth, New Hampshire, 1798–1837*, Pg. 147
36. A rather telling example of the favorable working conditions found in Dockum's shop can be found in Samuel Dockum's business records. Over a six-year period starting in 1834, all but one of the thirteen or fourteen craftsmen who had begun working in his shop were still to be found on the job, an incredibly high retention rate for that trade at that time in American history. [Source: Johanna McBrien, *The Furniture Craft Community of Portsmouth, New Hampshire, 1798–1837*, Pg. 146]
37. Moses Victor Safford, *The Magazine, Antiques*, Pg. 102
38. Jack Larkin, *The Reshaping of Everyday Life, 1790–1840*. (New York: Harper and Row, Publishers, Inc., 1988), Pg. 222
39. M. V. Brewington, *Shipcarvers of North America*, Pg. 72
40. *Catalogue of the Officers and Students of the New Hampton Literary and Biblical Institution at New Hampton, NHNH, for the Academical Year 1857–8*. Jones & Cogswell,—Printers, Concord, NHNH, 1858
41. William Hurlin, *The Academic and Theological Institution, New Hampton, NHNH*. The *Granite Monthly*, January, 1900, Pg. 17
42. William Hurlin, *The Academic and Theological Institution, New Hampton, NHNH*, Pg. 19
43. William Hurlin, *The Academic and Theological Institution, New Hampton, NHNH*, Pg. 18
44. *Catalogue of the Officers and Students of the New Hampton Literary and Biblical Institution at New Hampton, NHNH, for the Academical Year 1857–8*.
45. *History of the New Hampton Institution*. Collection of the Gordon-Nash Library, New Hampton, NH, Pg. 143
46. *Constitution of the New Hampton Literary and Biblical Institution*. Collection of the Gordon-Nash Library, New Hampton, NH
47. *History of the New Hampton Institution*, Pgs. 130–131; *Catalogue of the Officers and Students of the New Hampton Literary and Biblical Institution at New Hampton, NH, for the Academical Year 1857–8*.
48. William Hurlin, *The Academic and Theological Institution, New Hampton, NH*, Pgs. 27, 29
49. *The Portsmouth Herald*, December 17, 1927; Moses Victor Safford, *JHB*, Pg. 7
50. William Hurlin, *The Academic and Theological Institution, New Hampton, NH*, Pg. 26. Within this document it is stated: "Especially was it [the New Hampton Literary and Biblical Institution] distinguished for religious revivals every year. Such was its reputation in this respect, that throughout New Hampshire and other states,

parents sent their children there, hoping and expecting they would be converted and become Christians."

51. *History of the New Hampton Institution*, Pgs. 136–137
52. Source: Interview with Joseph W. P. Frost, Kittery Point local historian, as quoted within Yvonne Brault Smith, *John Haley Bellamy, Carver of Eagles*, Pg. 11
53. *Catalogue of the Officers and Students of the New Hampton Literary and Biblical Institution at New Hampton, NH, for the Academical Year 1857–8.*
54. Personal correspondence from John Haley Bellamy to Charles Gerrish Bellamy, April 29, 1859, collection of Dyer Library, Saco, ME
55. *Catalogue of the Officers and Students of the New Hampton Literary and Biblical Institution at New Hampton, NH, for the Academical Year 1857–8*
56. Personal correspondence from John Haley Bellamy to Charles Gerrish Bellamy, April 29, 1859, collection of Dyer Library, Saco, ME
57. *Catalogue of the Officers and Students of the New Hampton Literary and Biblical Institution at New Hampton, NH, for the Academical Year 1857–8*. Studying to become a teacher, Catherine Bellamy attended the school's female department, which "ranked among the best and most noted schools for young ladies in all the country…" [Source: William Hurlin, *The Academic and Theological Institution, New Hampton, NH*, Pg. 26]

Lancaster 3

1. M. V. Brewington, *Shipcarvers of North America*, Pg. 4. Prior to that year figureheads for American colonial vessels had been produced in England.
2. M. V. Brewington, *Shipcarvers of North America*, Pgs. 6, 8
3. Erwin O Christensen, *Early American Wood Carving*, Pg. 19
4. M. V. Brewington, *Shipcarvers of North America*, Pg. 57
5. M. V. Brewington, *Shipcarvers of North America*, Pg. 57
6. Pauline A. Pinckney, *American Figureheads and Their Carvers*, Pg. 160
7. M. V. Brewington, *Shipcarvers of North America*, Pg. 72
8. Pauline A. Pinckney, *American Figureheads and Their Carvers*, Pg. 159

Chapter 3

1. Yvonne Brault Smith, *John Haley Bellamy, Carver of Eagles*, Pg. 7
2. Moses Victor Safford, *The Magazine, Antiques*, Pgs. 103, 104
3. Xerox copy of personal correspondence from Charles Gerrish Bellamy to John Haley Bellamy, March 23, 1872, collection of Rice Public Library, Kittery, ME
4. Personal correspondence from John Haley Bellamy to Charles Gerrish Bellamy, November 10, 1873 (Acc# 1949.214), collection of the Farnsworth Art Museum and Wyeth Center
5. Phoenix Insurance Company of Hartford, Connecticut, Policy Number 192, December 19, 1869, collection of Dyer Library, Saco, ME. It must be noted that this insurance policy has in the past been inexplicably mistaken for an annuity taken out by a relative of Bellamy's intended to support him in his later years. [See Yvonne Brault Smith, *John Haley Bellamy, Carver of Eagles*, Pg. 97, footnote #2, Chapter VI.] Policy Number 192 clearly makes no mention of any annuity payment being made to John Haley Bellamy, especially as he and his brother Charles Carroll Bellamy are listed as the two individuals who initially took out the policy, nothing more than a $3,000 policy taken out expressly for the purposes of insuring the Pepperrell Mansion "AGAINST LOSS OR DAMAGE BY FIRE." John Haley Bellamy was in fact removed from the list of beneficiaries; it was ultimately listed that Mrs. Fanny Bellamy, Elisha P. Bellamy, and Charles C. Bellamy should be the policy's recipients.
6. Personal correspondence from John Haley Bellamy to Charles Gerrish Bellamy, January 25, 1874 (Acc# 1949.162), collection of the Farnsworth Art Museum and Wyeth Center
7. Personal correspondence from John Haley Bellamy to Charles Gerrish Bellamy, September 21, 1873 (Acc# 1949.164), collection of the Farnsworth Art Museum and Wyeth Center
8. Personal correspondence from John Haley Bellamy to Charles Gerrish Bellamy, November 10, 1873 (Acc# 1949.214), collection of the Farnsworth Art Museum and Wyeth Center
9. Personal correspondence from Charles Gerrish Bellamy to John Haley Bellamy, February 2, 1874 (Acc# 1949.59), collection of the Farnsworth Art Museum and Wyeth Center
10. Yvonne Brault Smith, *John Haley Bellamy, Carver of Eagles*, Pgs. 7, 14; Mitchell, Horace, and Jeanette: "Bellamy's Eagles." *American Mercury*, September, 1959, Pgs. 144–145
11. Personal correspondence from John Haley Bellamy to Charles Gerrish Bellamy, February 8, 1874 (Acc# 1949.16), collection of the Farnsworth Art Museum and Wyeth Center

12. Personal correspondence from John Haley Bellamy to Charles Gerrish Bellamy, September 21, 1873 (Acc# 1949.164), collection of the Farnsworth Art Museum and Wyeth Center
13. Personal correspondence from John Haley Bellamy to Charles Gerrish Bellamy, September 21, 1873 (Acc# 1949.164), collection of the Farnsworth Art Museum and Wyeth Center. It is interesting to note that Bellamy was viewed by his neighbors as an authority on the subject of what a rooster was emblematic of. This may be seen as further verification that Bellamy was considered by his peers as an authority on heraldry, as claimed by Moses Victor Safford.
14. Yvonne Brault Smith, *John Haley Bellamy, Carver of Eagles*, Pg. 13
15. The evidence that suggests that Bellamy was working as a woodcarver in Portsmouth, New Hampshire, in January 1859 comes from a note pasted on the back of a pine cat head sculpture by Bellamy, presently within the collection of Old Sturbridge Village (acc.# 20.16.4). The note reads, "This piece was carved by John H. Bellamy, at 77 Daniel Street, January 1859." As it is now known that Bellamy was indeed in business as a woodcarver on Daniel Street in Portsmouth that year, the date ascribed to this piece certainly bears a degree of legitimacy. Yet it must be noted that the address mentioned in the note appears incorrect, as we know Bellamy to have been in studio at 17 Daniel Street, not 77. Furthermore, Bellamy is not mentioned as having been in business until June of that year. It is thus the opinion of the author that if Bellamy was indeed working in Portsmouth as a woodcarver in January 1859, it was probably not as a self-employed woodcarver but rather as a journeyman for his former teacher Samuel Dockum, whose business was located at 17 Daniel Street at that time. Moreover, it was probably a brief stint, done only between terms at the New Hampton Institute.
16. Personal correspondence from John Haley Bellamy to Charles Gerrish Bellamy, April 29, 1859, collection of Dyer Library, Saco, ME. This document by its very existence disproves previous statements regarding Bellamy's tenure at that school, specifically the claim that "He [Bellamy] studied at New Hampton Institute for only one semester." See Yvonne Brault Smith, *John Haley Bellamy, Carver of Eagles*, Pg. 18.
17. *Catalogue of the Officers and Students of the New Hampton Literary and Biblical Institution at New Hampton, NH, for the Academical Year 1857–8*
18. By January 1853, Dockum had been forced to relocate his business from Portsmouth's Market Street, the very epicenter of the port's commerce, to nearby Daniel Street. [Source: *Portsmouth City Directory*, 1853] Located just "a few doors east of the Market" at the address of 17 Daniel Street See Johanna McBrien, *The Furniture Craft Community of Portsmouth, New Hampshire, 1798–1837*, Pg. 128, Dockum would be found at this address eleven years later, working under the title of "Carver and Cabinet Maker" during his final year of business. *Portsmouth City Directory*, 1864; R.W. Gerald D. Foss, *Samuel M. Dockum, Cabinetmaker*.
19. *Portsmouth Times*, June 18, 1859
20. See Moses Victor Safford, *The Magazine, Antiques*, Pg. 104; Philip M. Isaacson, *John H. Bellamy* (essay) within *Versatility – Yankee Style: The Cultural Diversity of Rev. Jonathan Fisher, John H. Bellamy and the Hardy Family*. Pg. 20; Yvonne Brault Smith, *John Haley Bellamy, Carver of Eagles*, Pg. 20. It must be noted that there is no personal or business listing for Bellamy at the 17 Daniel Street address in the *Portsmouth City Directory* throughout the 1860s, 70s, 80s, and 90s. As well, no Bellamy family personal correspondences make reference to John Haley Bellamy being in business at this address during these decades.
21. Philip M. Isaacson, *John H. Bellamy*. (essay) within *Versatility – Yankee Style: The Cultural Diversity of Rev. Jonathan Fisher, John H. Bellamy and the Hardy Family*, Pg. 20
22. Moses Victor Safford, *The Magazine, Antiques*, Pg. 103
23. *Portsmouth Herald*, April 6, 1914
24. *Portsmouth Times*, April 6, 1914
25. Typewritten copy of unidentified, undated obituary for John Haley Bellamy (Acc# 1949.217), collection of the Farnsworth Art Museum and Wyeth Center
26. Personal correspondence from John Haley Bellamy to Charles Gerrish Bellamy, November 10, 1873 (Acc# 1949.214), collection of the Farnsworth Art Museum and Wyeth Center
27. *The Heritage of Woodcarving: An Exhibition of Decorative Arts by John Haley Bellamy, Master Ship Carver and Seven Rockland, ME, Carvers, July 18 through September 7, 1975*. The William A. Farnsworth Library and Art Museum, 1975, Rockland, ME, Pg. 5
28. We know this piece to have been crafted by Bellamy in January 1859 from a note pasted on its back that reads: "This piece was carved by John H. Bellamy, at 77 Daniel Street, January 1859." [Source: E-mail communication with Tom Kelleher, curator, Old Sturbridge Village, April 1, 2008]. Some past authors on Bellamy have chosen to disregard this information and have instead elected to assign a date of about 1873 to this particular piece. [See Yvonne Brault Smith, *John Haley Bellamy, Carver of Eagles*, Pg. 85]
29. Erwin O Christensen, *Early American Wood Carving*, Pgs. 80–81. Though some have speculated that by carving this cat head in the shape of a lion that Bellamy was engaging in a pun of sorts, the fact is the lion motif upon a cat head was the convention of the time.

Lions were the sole design to be featured upon cat heads until the late nineteenth century, when carvers finally began to experiment with different motifs.

30. M. V. Brewington, *Shipcarvers of North America*, Pg. 71
31. Pauline A. Pinckney, *American Figureheads and Their Carvers*, Pg. 30
32. Pauline A. Pinckney, *American Figureheads And Their Carvers*, Pg. 30
33. Johanna McBrien, *The Furniture Craft Community of Portsmouth, New Hampshire, 1798–1837*, Pg. 142
34. M. V. Brewington, *Shipcarvers of North America*, Pg. 71
35. Personal correspondence from John Haley Bellamy to Frederick Lewis, October 6, 1859 (Acc# 1949.68), collection of the Farnsworth Art Museum and Wyeth Center
36. *Portsmouth Herald*, April 7, 1914
37. http://www.medfordhistorical.org/shipbuilding.php
38. 1860 US Census: *Schedule 1.– Free Inhabitants in Kittery in the County of York State of Maine*, June 1, 1860, Pg. 6
39. *Portsmouth Herald*, April 6, 1914
40. Richard E. Winslow III, *"Do Your Job!" An Illustrated Bicentennial History of the Portsmouth Naval Shipyard, 1800–2000*, Pgs. 20–21
41. http://www.medfordhistorical.org/shipbuilding.php
42. Frederick R. Black and Edwin C. Bearss, *The Charlestown Navy Yard, 1842–1890*, Pg. 203
43. 1860 US Census: *Schedule 1.– Free Inhabitants in Kittery in the County of York State of Maine*, June 1, 1860, Pg. 6; Moses Victor Safford, *Bellamy Genealogy*.
44. Letters of Incorporation penned by Charles Gerrish Bellamy, January 20, 1863, collection of Dyer Library, Saco, ME
45. Source: Interview with Wayne Manson, curator, Kittery Naval and Historical Museum, September 12, 2008. Most of the Kittery recipients of the Congressional Medal of Honor during the American Civil War controversially received their awards for service in the 27th Maine, merely for volunteering to defend Washington, DC, past the term of their enlistment.
46. Receipt of ten dollars from Charles C. Bellamy for filling the Town's Quota under the last call of the President, March 27, 1865, collection of Dyer Library, Saco, ME; Receipt of ten dollars from Elisha Bellamy for filling the Town's Quota under the last call of the President, March 27, 1865, collection of Dyer Library, Saco, ME
47. Frederick R. Black and Edwin C. Bearss, *The Charlestown Navy Yard, 1842–1890*, Pg. 202
48. *The Charlestown Directory*, 1864
49. M. V. Brewington, *Shipcarvers of North America*, Pg. 72
50. *Boston, Charlestown Navy Yard, 1800–1842, historic resource study, volume I of II.* Boston National Historical Park, Massachusetts, October, 1984, Pgs. 15, 111
51. Frederick R. Black and Edwin C. Bearss, *The Charlestown Navy Yard, 1842–1890*, Pg. 177
52. Frederick R. Black and Edwin C. Bearss, *The Charlestown Navy Yard, 1842–1890*, Pgs. 150, 160, 167
53. Frederick R. Black and Edwin C. Bearss, *The Charlestown Navy Yard, 1842–1890*, Pg. 179
54. Frederick R. Black and Edwin C. Bearss, *The Charlestown Navy Yard, 1842–1890*, Pg. 198
55. Frederick R. Black and Edwin C. Bearss, *The Charlestown Navy Yard, 1842–1890*, Pg. 198
56. Frederick R. Black and Edwin C. Bearss, *The Charlestown Navy Yard, 1842–1890*, Pgs. 201, 202
57. Frederick R. Black and Edwin C. Bearss, *The Charlestown Navy Yard, 1842–1890*, Pg. 201

Lancaster 4

1. Pauline A. Pinckney, *American Figureheads and Their Carvers*, Pgs. 159–160
2. Philip M. Isaacson, *The American Eagle*, Pgs. V, 6
3. Philip M. Isaacson, *The American Eagle*, Pg. 2
4. Philip M. Isaacson, *The American Eagle*, Pg. 2
5. Philip M. Isaacson, *The American Eagle*, Pg. 19
6. Philip M. Isaacson, *The American Eagle*, Pg. 48
7. Philip M. Isaacson, *The American Eagle*, Pg. 63
8. Philip M. Isaacson, *The American Eagle*, Pg. 138

Chapter 4

1. Frederick R. Black and Edwin C. Bearss, *The Charlestown Navy Yard, 1842–1890*, Pgs. 244, 245
2. Frederick R. Black and Edwin C. Bearss, *The Charlestown Navy Yard, 1842–1890*, Pg. 210
3. http://www.history.navy.mil/photos/sh-usn/usnsh-w/wabash.htm
4. Frederick R. Black and Edwin C. Bearss, *The Charlestown Navy Yard, 1842–1890*, Pg. 219
5. M. V. Brewington, *Shipcarvers of North America*. Barre Publishing Company, Barre, Massachusetts, 1962, Pg. 88

6. Pauline A. Pinckney, *American Figureheads and Their Carvers*. W.W. Norton & Company, Inc., New York, 1940, Pg. 42
7. Moses Victor Safford, *The Magazine, Antiques*, Pg. 103
8. *Portsmouth Herald*, April 6, 1914
9. *Portsmouth Herald*, April 7, 1914
10. *Portsmouth Times*, April 8, 1914
11. Frederick R. Black and Edwin C. Bearss, *The Charlestown Navy Yard, 1842–1890*, Pg. 247
12. Horace Mitchell, "Eagles with Feelings," Pg. 30
13. Personal correspondence from Elisha Bellamy to Charles Carroll Bellamy, April 2, 1866, collection of Dyer Library, Saco, ME
14. Personal correspondence from Elisha Bellamy to Charles Carroll Bellamy, April 2, 1866, collection of Dyer Library, Saco, ME
15. Declaration of intent to submit a patent to the US Patent Office penned by John Haley Bellamy, August 8, 1868, collection of the Phillips Library, Peabody Essex Museum
16. Description of a hauling device invention penned by John Haley Bellamy, n.d. (Acc# 1949.164), collection of the Farnsworth Art Museum and Wyeth Center. In her book *John Haley Bellamy, Carver of Eagles*, Yvonne Brault Smith confuses this invention with another of Bellamy's creations, namely a hauling device crafted by the artist when employed at the Portsmouth Navy Yard in the late 1870s/ early 1880s specifically to draw the USS *Plymouth* out of the water. See Yvonne Brault Smith, *John Haley Bellamy, Carver of Eagles*, Pgs. 29–30. As there is roughly a decade between the patenting of the hauling device mentioned in this patent application and his envisioning a hauling device for the *Plymouth*, and as there is no record that Bellamy ever attempted to get his *Plymouth* hauling device patented, it becomes obvious that Smith has taken two similar but separate Bellamy inventions and combined them into one.
17. Declaration of intent to submit a patent to the US Patent Office penned by John Haley Bellamy, January, 1873, collection of the Phillips Library, Peabody Essex Museum
18. Yvonne Brault Smith, *John Haley Bellamy, Carver of Eagles*, Pg. 14
19. Moses Victor Safford, *The Magazine, Antiques*, Pg. 103
20. Legal correspondence from John Haley Bellamy to Lois M. Faxon, May 31, 1869, collection of the Phillips Library, Peabody Essex Museum
21. Legal correspondence from Lois M. Faxon to John Haley Bellamy, November 16, 1869, collection of the Phillips Library, Peabody Essex Museum
22. *Masonic Symbols in American Decorative Arts*. Exhibition catalog, Scottish Rite Masonic Museum Of Our National Heritage, Lexington, Massachusetts, 1976, Pg. 31
23. *Masonic Symbols in American Decorative Arts*, Pgs. 17, 31–32
24. *Masonic Symbols in American Decorative Arts*, Pg. 39
25. *Masonic Symbols in American Decorative Arts*, Pg. 39
26. *Masonic Symbols in American Decorative Arts*, Pg. 40
27. *Masonic Symbols in American Decorative Arts*, Pg. 19
28. E-mail communication with Aimee E. Newell, Curator of Collections, National Heritage Museum, April 2, 2008
29. Yvonne Brault Smith, *John Haley Bellamy, Carver of Eagles*, Pg. 18
30. *Masonic Symbols in American Decorative Arts*, Pg. 42
31. Typewritten copy of personal correspondence from John Haley Bellamy to Charles Gerrish Bellamy, March 3, 1866 (Acc# 1949.217), collection of the Farnsworth Art Museum and Wyeth Center
32. Typewritten copy of personal correspondence from John Haley Bellamy to Charles Gerrish Bellamy, August 8, 1868, collection of Dyer Library, Saco, ME
33. Statement of half interest in the Masonic frame inventions of John Haley Bellamy penned by D.A. Titcomb. November 2, 1868, collection of the Phillips Library, Peabody Essex Museum
34. *Masonic Symbols in American Decorative Arts*, Pg. 39
35. Typewritten copy of personal correspondence from John Haley Bellamy to Charles Gerrish Bellamy, March 3, 1866 (Acc# 1949.217), collection of the Farnsworth Art Museum and Wyeth Center
36. Business agreement between John Haley Bellamy, Eldridge Walcott and Jonas C. Young, May 21, 1870, collection of the Phillips Library, Peabody Essex Museum
37. Business agreement between John Haley Bellamy and Henry C. Merriam, November 26, 1868, collection of the Phillips Library, Peabody Essex Museum
38. Business agreement between John Haley Bellamy, Eldridge Walcott and Jonas C. Young, May 30, 1870, collection of the Phillips Library, Peabody Essex Museum
39. Business agreement between John Haley Bellamy and David A. Titcomb, July 30, 1866, collection of the Phillips Library, Peabody Essex Museum
40. Business agreement between John Haley Bellamy and David A. Titcomb, July 30, 1866
41. *The Charlestown Directory*, 1868, and *Bellamy Banter: The Life and Work of John Haley Bellamy*. Piscataqua Decorative Arts Society, Vol. III, 2007–2009 Lecture Series, 2011, Pg. 4

42. Business agreement between John Haley Bellamy and David A. Titcomb, July 30, 1866
43. Business agreement between John Haley Bellamy and David A. Titcomb, July 30, 1866
44. Business agreement between John Haley Bellamy and David A. Titcomb, July 13, 1868, collection of the Phillips Library, Peabody Essex Museum
45. Personal correspondence from John Haley Bellamy to Charles Gerrish Bellamy, March 11, 1868 (Acc# 1949.251), collection of the Farnsworth Art Museum and Wyeth Center
46. Statement of half interest in the Masonic frame inventions of John Haley Bellamy penned by D.A. Titcomb. November 2, 1868, collection of the Phillips Library, Peabody Essex Museum
47. *Massachusetts*, Vol. 79, p. 233, R.G. Dun & Co. Collection, Baker Library Historical Collections, Harvard Business School.
48. Yvonne Brault Smith, *John Haley Bellamy, Carver of Eagles*, Pg. 18. Smith makes the claim that Bellamy patented clock cases with Grand Army of the Republic and Knights of Columbus insignia without citing any source or example of such work to support her claim. In addition, during his research in preparation for writing this book the author was unable to locate any source, citation, or actual example of such work to support Ms. Smith's claim. Given the popularity of the GAR organization at that time in history and Bellamy's demonstrated business sense in identifying lucrative markets for his clock cases, the author has elected to include Ms. Smith's claim that Bellamy made GAR-themed clock cases given the reasonable likelihood that indeed he did, all while trusting in the veracity of Ms. Smith's words. Regarding the claim by Ms. Smith that Bellamy made Knights of Columbus-themed clock cases, it is the author's opinion that Ms. Smith may have mistakenly assumed that his Papal Clock—a design that was intended for being marketed and sold to the Catholic community at large—was meant only for the Knights of Columbus.
49. Handbill advertisement for the Papal Clock, (Acc# 1949.253) collection of the Farnsworth Art Museum and Wyeth Center
50. Philip M. Isaacson, *The American Eagle*, Pgs. 64–65

Lancaster 5

1. *Portsmouth Times*, December 12, 1879
2. *Portsmouth Herald*, April 7, 1914
3. See Sessions, Ralph: *The Shipcarver's Art: Figureheads and Cigar-Store Indians in Nineteenth-Century America*. Princeton University Press, Princeton and Oxford, 2005, Pg. 73, for a superb accounting of how nineteenth-century woodcarvers crafted ship's figureheads.
4. MVS: *JHB*, Pg. 17. It must be mentioned that in her book *John Haley Bellamy, Carver of Eagles*, Yvonne Brault Smith confuses this invention with another of Bellamy's creations, namely a generic hauling device crafted by the artist for use on land and sea. See Yvonne Brault Smith, *John Haley Bellamy, Carver of Eagles*, Pgs. 29–30.

Chapter 5

1. Yvonne Brault Smith, *John Haley Bellamy, Carver of Eagles*, Pg. 18
2. List of Bellamy Family Birth and Death Dates, Pg. 5 (Acc# 1949.185), collection of the Farnsworth Art Museum and Wyeth Center
3. MVS: *Eagle,* Pg. 4. Safford's words directly disprove the unfounded claim that Charles Gerrish Bellamy entered into bankruptcy in the mid-1890s, when he was an elderly man in his eighties. [See Yvonne Brault Smith, *John Haley Bellamy, Carver of Eagles*, Pg. 49]
4. *Portsmouth Herald*, January 14, 1899
5. Personal correspondence from Charles Gerrish Bellamy to John Haley Bellamy, July 20, 1873, collection of Rice Public Library, Kittery, ME
6. Phoenix Insurance Company of Hartford, Connecticut, Policy Number 192, December 19, 1869, collection of Dyer Library, Saco, ME
7. Personal correspondence from John Haley Bellamy to Charles Gerrish Bellamy, November 10, 1873 (Acc# 1949.214), collection of the Farnsworth Art Museum and Wyeth Center
8. Richard E. Winslow III, *"Do Your Job!" An Illustrated Bicentennial History of the Portsmouth Naval Shipyard, 1800–2000*, Pgs. 2, 24
9. Typewritten copy of personal correspondence from Frances Keene Bellamy to Charles Gerrish Bellamy, September 25, 1873, collection of Dyer Library, Saco, ME
10. *Portsmouth Herald*, April 6, 1914
11. Personal correspondence from John Haley Bellamy to Charles Gerrish Bellamy, November 10, 1873, (Acc# 1949.214) collection of the Farnsworth Art Museum and Wyeth Center
12. Moses Victor Safford, *The Magazine, Antiques*, Pg. 103
13. *Portsmouth Herald*, April 6, 1914
14. Richard E. Winslow III, *"Do Your Job!" An Illustrated Bicentennial History of the Portsmouth Naval Shipyard, 1800–2000*, Pgs. 40, 42

15. http.//www.mariner.org/exhibitions/highlights/figurehead_bear.php. 2008
16. Personal correspondence from John Haley Bellamy to Charles Gerrish Bellamy, November 10, 1873 (Acc# 1949.214), collection of the Farnsworth Art Museum and Wyeth Center
17. Moses Victor Safford, *The Magazine, Antiques*, Pg. 103
18. Moses Victor Safford, *The Magazine, Antiques*, Pg. 103
19. Moses Victor Safford, *The Magazine, Antiques*, Pg. 103
20. MVS: *JHB*, Pg. 16. As is somewhat typical of Safford's recollections of John Haley Bellamy, contradictions exist within his testimonies. While he claims in this passage that Bellamy would have gone home had he taken ill, he would also state that "He [Bellamy] would not be likely to lie down in his workshop unless taken sick" [See MVS: *Eagle*, Pgs. 10–11], thus giving the impression that Bellamy could have lied down in his shop when ill, rather than going home. Furthermore, while he states here that the *Lancaster* eagle figurehead was never assembled together in Bellamy's shop, he would inexplicably contradict himself by stating: "I recall seeing Bellamy on a platform directing the bolting of sections of the eagle together on the bow. This impressed me because I had previously seen the entire eagle assembled and held together somehow in the mold loft." [See MVS: *Eagle*, Pg. 13.]
21. Frederick R. Black and Edwin C. Bearss, *The Charlestown Navy Yard, 1842–1890*, Pg. 203
22. Frederick R. Black and Edwin C. Bearss, *The Charlestown Navy Yard, 1842–1890*, Pg. 247
23. *Navy Yard Orders, Form 5, Trade Card, Schedule "B," requisition number 817–119, John H. Bellamy*. National Personnel Records Center, Civilian Personnel Records
24. David A. Wasson, "The Silent Pilots." *The Outlook*, Volume CIX, January 27, 1915, Pg. 213. As this story was penned by the son of Bellamy's good friend George S. Wasson, there may be a certain grain of truth present within the tale which, with the passing of the years and retellings, took on the fantastical, comical dimensions found within David Wasson's retelling. It must be further noted that David A. Wasson was not yet born when the purported assault by Bellamy against the Navy officer took place, and that he was a mere child during the majority of years his father spent in friendship with and as a neighbor of John Haley Bellamy. (David A. Wasson was born in 1887.)
25. Personal correspondence from John Haley Bellamy to Charles Gerrish Bellamy, September 21, 1873 (Acc# 1949.164), collection of the Farnsworth Art Museum and Wyeth Center
26. Personal correspondence from John Haley Bellamy to Charles Gerrish Bellamy, September 21, 1873

Lancaster 6

1. *Extracts from the Daily Log Book*, August 21, 1881
2. Philip M. Isaacson, *The American Eagle*, Pg. 83
3. *Portsmouth Daily Chronicle*, September 13, 1881.
4. Philip M. Isaacson, *The American Eagle*, Pg. 8

Chapter 6

1. Personal correspondence from John Haley Bellamy to Charles Gerrish Bellamy, September 17, 1872 (Acc# 1949.155), collection of the Farnsworth Art Museum and Wyeth Center
2. Personal correspondence from John Haley Bellamy to Charles Carroll Bellamy, December 12, 1872 (Acc# 1949.248), collection of the Farnsworth Art Museum and Wyeth Center
3. Source: Interview with Steven Estes, Kittery Point local historian, July 10, 2008
4. Personal correspondence from John Haley Bellamy to Charles Gerrish Bellamy, October 2, 1872 (Acc# 1949.168), collection of the Farnsworth Art Museum and Wyeth Center
5. Personal correspondence from John Haley Bellamy to Charles Carroll Bellamy, December 12, 1872 (Acc# 1949.248), collection of the Farnsworth Art Museum and Wyeth Center
6. Philip M. Isaacson, *The American Eagle*, Pg. 118. It should be noted that Isaacson mistakenly gives a 1890s date for the Appledore burning down.
7. Philip M. Isaacson, *The American Eagle*, Pg. 118
8. Personal correspondence from John Haley Bellamy to Charles Gerrish Bellamy, March 16, 1873 (Acc# 1949.161), collection of the Farnsworth Art Museum and Wyeth Center
9. The placement of the Bellamy eagle that graced the portico over the entrance to the Newfields town hall may only date to the 1920s. According to William C. Tebo, a former town selectman of Newfields, the Bellamy eagle may have been donated by a local resident to adorn the town hall sometime between the years 1927–28, as the town hall was then undergoing considerable renovation and construction. [Source: Interview with William C. Tebo, former town selectman, Newfields, New Hampshire, 11/13/2008] Yet given the peculiar dimensions and design of this particular Bellamy eagle,

combined with its similarities to other known commissions Bellamy received from local governments and businesses, the fact that Mr. Tebo could not definitely declare that the eagle had been installed in 1927–28, and the absence of any documentary evidence that can declare a definite date for its installation at present, the author withholds stating a definite verdict as to when the installation of the Bellamy eagle atop the Newfields, New Hampshire town hall took place, and has chosen to include its mention here within the main text as a possible example of the type of work Bellamy was doing for local governments during the 1870s to 1890s.

10. Personal correspondence from John Haley Bellamy and family to Charles Gerrish Bellamy, October 13, 1872 (Acc# 1949.57), collection of the Farnsworth Art Museum and Wyeth Center
11. Horace and Jeanette Mitchell, "Bellamy's Eagles," Pg. 146
12. Horace and Jeanette Mitchell, "Bellamy's Eagles," Pg. 146
13. Source: Interview with Arthur Swanson, woodcarver, restorer, and noted producer of Bellamy reproductions, July 20, 2008
14. Philip M. Isaacson, *The American Eagle*, Pg. 118
15. Source: Interview with Arthur Swanson, July 20, 2008
16. Horace Mitchell, "Eagles with Feelings," Pg. 29
17. Yvonne Brault Smith, *John Haley Bellamy, Carver of Eagles*, Pg. 38
18. Personal correspondence from John Haley Bellamy to Charles Gerrish Bellamy, January 25, 1874 (Acc# 1949.162), collection of the Farnsworth Art Museum and Wyeth Center
19. Source: Interview with Steven Estes, Kittery Point local historian, July 10, 2008
20. Personal correspondence from John Haley Bellamy to Charles Gerrish Bellamy, January 25, 1874
21. Personal correspondence from John Haley Bellamy to Charles Gerrish Bellamy, January 25, 1874
22. Personal correspondence from Charles Gerrish Bellamy to John Haley Bellamy, February 2, 1874 (Acc# 1949.59), collection of the Farnsworth Art Museum and Wyeth Center
23. Personal correspondence from Charles Gerrish Bellamy to John Haley Bellamy, January 31, 18[?], collection of Rice Public Library, Kittery, ME
24. Personal correspondence from Charles Gerrish Bellamy to John Haley Bellamy, February 26, 1873, collection of Rice Public Library, Kittery, ME
25. Personal correspondence from Charles Gerrish Bellamy to John Haley Bellamy, January 31, 18[?],
26. Moses Victor Safford, *The Magazine, Antiques*, Pg. 103
27. Source: Interview with Wayne Manson, curator, Kittery Naval and Historical Museum, September 12, 2008.
28. To the author's knowledge, as of the time of this publication no Bellamy eagle sporting the saying "Happy Thanksgiving" has been discovered. The source for claiming that Bellamy produced eagles sporting this aphorism is a personal correspondence dating from September of 1872, wherein Bellamy informs his father, "All is well and moving on as usual. I am still in the shop making Eagles have some very fine birds on hand – for Thanksgiving." [Source: Personal correspondence from John Haley Bellamy to Charles Gerrish Bellamy, September 17, 1872, (Acc# 1949.155) collection of the Farnsworth Art Museum and Wyeth Center]
29. Moses Victor Safford, *The Magazine, Antiques*, Pg. 103
30. Horace Mitchell, "Eagles with Feelings," Pg. 33
31. *Portsmouth Daily Chronicle*, September 2, 1885. It should be noted that local legends gleaned by the author allude to Bellamy having filled out orders for his classic two foot-long eagles from outside the United States, including the West Indies, where Bellamy is rumored to have once sent a shipment of 350 eagles, and Argentina, where a request for 1,400 eagles is said to have come in. While quite specific in their nature, the author has chosen not to include these claims within the main text due to an inability on his part to uncover any corroborating evidence that supports them as of the time of publication.
32. Moses Victor Safford, *The Magazine, Antiques*, Pg. 103
33. Personal correspondence from John Haley Bellamy to Charles Gerrish Bellamy, January 25, 1874 (Acc# 1949.162), collection of the Farnsworth Art Museum and Wyeth Center
34. Brighton, Ray: *Frank Jones King of the Alemakers*. Peter E. Randall, Publisher, Hampton, NH, 1976, Pg. 145
35. Moses Victor Safford, *The Magazine, Antiques*, Pg. 104
36. *The Heritage of Woodcarving: An Exhibition of Decorative Arts by John Haley Bellamy, Master Ship Carver and Seven Rockland, Maine, Carvers, July 18 through September 7, 1975*. The William A. Farnsworth Library and Art Museum, 1975, Rockland, ME, Pg. 5
37. Moses Victor Safford, *The Magazine, Antiques*, Pg. 104; Pg. 103
38. Moses Victor Safford, *The Magazine, Antiques*, Pg. 103
39. Source: E-mail communication with David Lintz, Director, Red Men Museum and Library, November 7, 2008.

Lancaster 7

1. *Extracts from the Daily Log Book,* September 8, 1881
2. http://www.history.navy.mil/danfs/l3/lancaster-i.htm
3. http://www.history.navy.mil/danfs/l3/lancaster-i.htm
4. http://www.history.navy.mil/danfs/l3/lancaster-i.htm
5. http://www.history.navy.mil/danfs/l3/lancaster-i.htm
6. Yvonne Brault Smith, *John Haley Bellamy, Carver of Eagles,* Pgs. 31–34

Chapter 7

1. Homer Eaton Keyes, *Bellamy's Style and Its Imitators. The Magazine, Antiques*, Vol. XXVII, No. 3, March, 1935, Pg. 106
2. Horace Mitchell, "Eagles with Feelings," Pg. 28
3. Horace and Jeanette Mitchell, "Bellamy's Eagles," Pg. 146
4. Horace Mitchell, "Eagles with Feelings," Pg. 28
5. Horace Mitchell, "Eagles with Feelings," Pg. 29
6. Horace and Jeanette Mitchell, "Bellamy's Eagles," Pg. 146
7. Ronald Bourgeault, Bellamy Banter: The Life and Work of John Haley Bellamy," Rux, Sandra and Volk, Joyce Geary (editors). *Piscataqua Decorative Arts Society*, Vol. III, 2007–2009 Lecture Series (2011). Pg.5; Bourgeault, Ronald "John Haley Bellamy 1836–1914" Northeast Auctions Marine & China Trade catalog, August 21, 2005, Pg. 178
8. Ronald Bourgeault, *John Haley Bellamy 1836–1914*, Pg. 178
9. Homer Eaton Keyes, *Bellamy's Style and Its Imitators*, Pg. 107
10. Homer Eaton Keyes, *Bellamy's Style and Its Imitators*, Pg. 107
11. Homer Eaton Keyes, *Bellamy's Style and Its Imitators*, Pg. 107. For actual photographic illustrations of two Spinney attempts at copying the classic Bellamy eagle (one attributed, one unattributed) this article is a superb point of reference.
12. Source: Garth's Auction Inc., Delaware, Ohio. Auction records for item number 230306041681, John Haley Bellamy eagle sold on November 29, 2008
13. Source: Interview with David Kaselaukas of Kittery Point, ME, November 29, 2008
14. Moses Victor Safford, *The Magazine, Antiques*, Pg. 103
15. Yvonne Brault Smith, *John Haley Bellamy, Carver of Eagles*, Pg. 70
16. Philip M. Isaacson, *The American Eagle*, Pg. 116
17. Philip M. Isaacson, *John H. Bellamy*. (essay) within *Versatility—Yankee Style: The Cultural Diversity of Rev. Jonathan Fisher, John H. Bellamy and the Hardy Family*. Pg. 20
18. Moses Victor Safford, *The Magazine, Antiques*, Pg. 103
19. There is at present disagreement in the literature of John Haley Bellamy as to the type of wood this sculpture is composed of. The William A. Farnsworth Library and Art Museum identifies it as being either maple or birch [Source: *Versatility – Yankee Style: The Cultural Diversity of Rev. Jonathan Fisher, John H. Bellamy and the Hardy Family*. Pg. 30], while it has also on occasion been cited as being crafted from walnut. [Source: Yvonne Brault Smith, *John Haley Bellamy, Carver of Eagles*, Pg. 85.]
20. Moses Victor Safford, *John Haley Bellamy, The Woodcarver*. Unpublished manuscript dated November 23, 1941, collection of the Mariner's Museum, Newport News, VA., Pg. 8
21. Richard M. Candee, "The New Colonials: Restoration and Remodeling of Old Buildings Along the Piscataqua," Pg. 52. A fascinating read documenting the "restoration" of Sparhawk Hall is Arthur Little's *Early American Interiors* (Boston 1878), a text that stands as the earliest American book to illustrate the renovation of this property, and which describes the mansion's famous east room/ front parlor as formerly being "in a state of dilapidation, most of the panels being gone, but the present owner has restored it fully."
22. *Portsmouth Herald*, October 11, 2007
23. Horace Mitchell, "Eagles with Feelings," Pg. 29
24. Horace and Jeanette Mitchell, "Bellamy's Eagles," Pg. 146
25. Richard M. Candee, "Sparhawk Hall," Pg. 53, in *A Noble and Dignified Stream: The Piscataqua Region in the Colonial Revival, 1860–1930*, Sarah Giffen and Kevin Murphy, eds., York: Old York Historical Soc., 1992
26. This inscription was first publicly stated in the pages of the the *Kittery Press*, Kittery, ME, February 6, 1953 edition, and rediscovered by Brian Cooper of Deschenes & Cooper Architectural Millwork, LLC, and Early New England Restorations, LLC, of Pawcatuck, Connecticut, in 2012, during removal from its second location
27. Ellen MacDonald Ward, "Only a Memory," *DownEast Magazine*, v. 39 (Feb. 1993) Pgs. 53–54; Richard M. Candee, "The New Colonials: Restoration and Remodeling of Old Buildings Along the Piscataqua," Pgs. 35–48; "Sparhawk Hall," Pgs. 52–55, in *A Noble and Dignified Stream: The Piscataqua Region in the Colonial Revival, 1860–1930*
28. *Department of Commerce and Labor Bureau of Navigation Thirty-Sixth Annual List of Merchant Vessels of the United States*. Government Printing Office, Washington, 1904, Pg. 70

29. Interestingly, in 1893 this same schooner was believed to have been lost; it was reported that she had been "probably blown out to sea" off Portland, ME, on February 20, 1893. [Source: *New York Times*, March 10, 1893.]
30. Philip M. Isaacson, "John H. Bellamy" (essay), within *Versatility – Yankee Style: The Cultural Diversity of Rev. Jonathan Fisher, John H. Bellamy and the Hardy Family*, Pg. 18

Lancaster 8

1. Yvonne Brault Smith, *John Haley Bellamy, Carver of Eagles*, Pgs. 31–34
2. Yvonne Brault Smith, *John Haley Bellamy, Carver of Eagles*, Pg. 33
3. Yvonne Brault Smith, *John Haley Bellamy, Carver of Eagles*, Pg. 34

Chapter 8

1. Horace and Jeanette Mitchell, "*Bellamy' Eagles*," Pg. 144
2. No source contemporary to Bellamy has yet to be uncovered that makes specific mention of Admiral George Dewey having been a guest at Bellamy's shop. The only sources for this claim at present are the writings of Horace Mitchell, who himself makes no mention of where he obtained this information. As Mitchell himself had no obvious or apparent connection to John Haley Bellamy beyond his having lived in Sparhawk Hall long after Bellamy's passing, it seems that this claim is anecdotal at best.
3. John D. Bardwell, *Images of America: Old Kittery*. Arcadia Publishing, an imprint of the Chalford Publishing Corporation, Dover, New Hampshire, 1995, Pg. 7
4. Two retired mariners that we know to have regularly been found in Bellamy's workshop were a Richard Seward, "a retired master mariner" who, as paymaster's steward on the USS *Commodore*, received the Congressional Medal of Honor for his bravery at Ship Island Sound, Louisiana, [See: Moses Victor Safford, *JHB*, Pg. 10] and retired shipmaster Captain John Lawrence. [See: Moses Victor Safford, *The Magazine, Antiques*, Pg. 105]
5. Moses Victor Safford, *The Magazine, Antiques*, Pg. 103
6. Moses Victor Safford, *The Magazine, Antiques*, Pg. 104
7. Moses Victor Safford, *The Magazine, Antiques*, Pg. 104
8. Moses Victor Safford, *The Magazine, Antiques*, Pgs. 104–105
9. Fannie Hardy Eckstorm, *George Savary Wasson, Artist and Writer, 1855–1932*. Essex Institute Historical Collections, LXXIX (1943), Pg. 50; Cecil D. Eby Jr., "DownEast with George Savary Wasson." *DownEast Magazine*, Vol. XIII, No. 7, April, 1967, Pg. 46
10. Fannie Hardy Eckstorm, *George Savary Wasson, Artist and Writer, 1855–1932*, Pg. 51
11. Cecil D. Eby Jr., "DownEast with George Savary Wasson," Pg. 46
12. Cecil D. Eby Jr., "DownEast with George Savary Wasson," Pg. 46
13. Introduction for unpublished typewritten manuscript concerning George Savary Wasson presented to the Boston Athenaeum in January of 1949, prepared by Walter Muir Whitehill, and Cecil D. Eby Jr., "DownEast with George Savary Wasson," Pg. 46
14. Francis W. Hatch, "'Wave Crest' Retires," *DownEast Magazine*, Vol. XV, No. 9, June, 1969, Pg. 34
15. Introduction for unpublished typewritten manuscript concerning George Savary Wasson presented to the Boston Athenaeum in January of 1949, prepared by Walter Muir Whitehill, and Fannie Hardy Eckstorm, *George Savary Wasson, Artist and Writer, 1855–1932*. Essex Institute Historical Collections, LXXIX (1943), Pg. 54
16. Introduction for unpublished typewritten manuscript concerning George Savary Wasson presented to the Boston Athenaeum in January of 1949, prepared by Walter Muir Whitehill
17. Cecil D. Eby Jr., "DownEast with George Savary Wasson," Pg. 47
18. Fannie Hardy Eckstorm, *George Savary Wasson, Artist and Writer, 1855–1932*, Pgs. 54–55
19. Fannie Hardy Eckstorm, *George Savary Wasson, Artist and Writer, 1855–1932*, Pg. 47
20. Cecil D. Eby Jr., "DownEast with George Savary Wasson," Pg. 43
21. Fannie Hardy Eckstorm, *George Savary Wasson, Artist and Writer, 1855–1932*, Pg. 47
22. Fannie Hardy Eckstorm, *George Savary Wasson, Artist and Writer, 1855–1932*, Pg. 58
23. Fannie Hardy Eckstorm, *George Savary Wasson, Artist and Writer, 1855–1932*, Pg. 53
24. Moses Victor Safford, *The Magazine, Antiques*, Pg. 103
25. List of books associated with the Bellamy family, n.d., collection of Dyer Library, Saco ME
26. Moses Victor Safford, *The Magazine, Antiques*, Pg. 103
27. Moses Victor Safford, *The Magazine, Antiques*, Pg. 104
28. Horace and Jeanette Mitchell, "Bellamy's Eagles," Pg. 146
29. Moses Victor Safford, *The Magazine, Antiques*, Pg. 104
30. John Haley Bellamy, Untitled, un poem (Acc# 1949.119), collection of the Farnsworth Art Museum and Wyeth Center

31. John Haley Bellamy, *So*. n.d. Courtesy of Wayne Manson, curator, Kittery Historical and Naval Society.
32. John Haley Bellamy, *Tomorrow*. Unpublished, un poem (Acc# 1949.96), collection of the Farnsworth Art Museum and Wyeth Center
33. Yvonne Brault Smith, *John Haley Bellamy, Carver of Eagles*, Pgs. 43–44
34. Personal correspondence from John Haley Bellamy to S. M. Burgess, Massachusetts Humane Society, November 26, 1900 (Acc# 1949.163), collection of the Farnsworth Art Museum and Wyeth Center
35. "Tourists, Attention" handbill advertisement, n.d. (Acc# 1949.113), collection of the Farnsworth Art Museum and Wyeth Center. It should be noted that past authors have attributed the opening of the Pepperrell Mansion to tourists as "a scheme to make money," devised solely by Charles Gerrish Bellamy. [See Yvonne Brault Smith, *John Haley Bellamy, Carver of Eagles*, Pg. 49] At present there is no known documentation that attributes this decision solely to Charles Gerrish Bellamy. From Bellamy family letters dating to around 1895, it is known that whomever happened to be home when a tourist called upon the Bellamy household for a tour would host them and guide them through. As with so many other ventures and activities engaged in by the Bellamy family, it would appear that the decision to open their home to tourists was a joint venture, one that was conceived and executed by more than one member of this family.
36. John Haley Bellamy, *Atropos*. n.d. (Acc# 1949.123), collection of the Farnsworth Art Museum and Wyeth Center

Lancaster 9

1. Yvonne Brault Smith, *John Haley Bellamy, Carver of Eagles*, Pg. 34. Smith asserts that the asking price for the figurehead was $2,500, a claim that is readily disproven when viewing the photograph of the *Lancaster* Eagle in the ship's chandlery published in this book.
2. Yvonne Brault Smith, *John Haley Bellamy, Carver of Eagles*, Pg. 34
3. Yvonne Brault Smith, *John Haley Bellamy, Carver of Eagles*, Pg. 34
4. Yvonne Brault Smith, *John Haley Bellamy, Carver of Eagles*, Pg. 34

Chapter 9

1. Moses Victor Safford, *The Magazine, Antiques*, Pg. 105
2. Moses Victor Safford, *Bellamy Genealogy*.
3. Moses Victor Safford, *Bellamy Genealogy*.
4. *Portsmouth Herald*, January 14, 1899
5. Department of Commerce and Labor–Bureau of the Census: *Thirteenth Census of the United States: 1910– Population, Supervisors District No. 1, Enumeration District No. 246, Kittery, Maine*, April 20, 1910, Sheet No. 3
6. Moses Victor Safford, *Bellamy Genealogy*.
7. John Haley Bellamy, *Yes*. n.d. Courtesy of Wayne Manson, curator, Kittery Historical and Naval Society. The Margery Bray referred to repeatedly in the poem is Margery Bray Pepperrell, wife of Colonel William Pepperrell, founder of the Pepperrell dynasty, who in her life was noted for being "very exemplary for unaffected piety and amiable virtues, especially her charity, her courteous affability, her prudence, meekness, patience, and her unweariedness in welldoing. She was not only a loving and discreet wife and tender parent, but a sincere friend to all her acquaintance." [Source: Everett S. Stackpole, *Old Kittery and Her Families*. Pg. 246]
8. Moses Victor Safford, *The Magazine, Antiques*, Pg. 105
9. *Portsmouth Herald*, January 4, 1907
10. *The Heritage of Woodcarving: An Exhibition of Decorative Arts by John Haley Bellamy, Master Ship Carver and Seven Rockland, ME, Carvers, July 18 through September 7, 1975*. The William A. Farnsworth Library and Art Museum, 1975, Rockland, ME, Pg. 5
11. *Navy Yard Orders, Form 5, Trade Card, Schedule "B," requisition number 817–119, John H. Bellamy*. National Personnel Records Center, Civilian Personnel Records
12. *Navy Yard Orders, Form 5, Trade Card, Schedule "B," requisition number 817–119, John H. Bellamy*. National Personnel Records Center, Civilian Personnel Records
13. Docket #54609: *Guardian Account, Schedule A, The Account of John D. Carty Guardian of John H. Bellamy of Kittery, in the County of York*. Probate Court Records, York County, ME, and Personal correspondences from D.N. Simson, Superintendent, The Equitable Life Assurance Society of the United States to John Haley Bellamy, January 20, 1909, and April 20, 1909, collection of Dyer Library, Saco, ME.
14. Personal correspondences from D. N. Simson, Superintendent, The Equitable Life Assurance Society of the United States to John Haley

Bellamy, January 20, 1909, and April 20, 1909, collection of Dyer Library, Saco, ME

15. Invoice from Houghton Mifflin & Co.'s Periodicals to John Haley Bellamy, December 6, 1906, collection of Dyer Library, Saco, ME
16. Yvonne Brault Smith, *John Haley Bellamy, Carver of Eagles*, Pgs. 57–58
17. Personal correspondence from Fanny Bellamy to Charles Gerrish Bellamy, September 5, c. 1870(?), (Acc# 1949.58) collection of the Farnsworth Art Museum and Wyeth Center
18. Record of Arrests for the City of Portsmouth, New Hampshire, 1887–1892, entry for October 30, 1891, Pg. 460, collection of the Portsmouth Athenaeum, and Record of Arrests for the City of Portsmouth, New Hampshire, 1901–1905, entry for December 11, 1902, Pg. 122, collection of the Portsmouth Athenaeum. There is no record of Bellamy having ever been arrested in his native Kittery for this offense, an unsurprising occurrence, as Kittery did not have a police department until the 1930s. [Source: Telephone interview with Edward F. Strong, police chief, Kittery, ME, 1/21/2009]
19. One other purported but to date unverified letter also concerning this aspect of the artist's life is claimed to exist. Said to have been penned by Fanny Bellamy to her husband Charles on September 28, 1870, this letter supposedly finds her complaining, "Paying for whiskey don't keep folks out of the Almshouse… if he has to have liquor, I must have bread and butter" and further laments, "If he could just let the liquor alone…" [Source: Yvonne Brault Smith, *John Haley Bellamy, Carver of Eagles*, Pg. 19.] Yet Ms. Smith fails to cite the repository wherein this letter is deposited. This failure to properly identify the letter's location, combined with the fact that no letter written to Charles Gerrish Bellamy bearing the date September 28, 1870, is registered within any known public collection of Bellamy family correspondences ultimately makes it impossible to quote this letter in its entirety or even establish whether or not it truly exists. Furthermore, the grammatical correctness found within the September 28, 1870, Fanny Bellamy quote makes it further suspect, as Fanny Bellamy reveals herself in her documented letters to have been quasi-illiterate, prone to phonetic, incorrect spelling and improper sentence structure on a sentence-by-sentence basis. Consequently, given these factors and the presence of numerous other discrepancies specifically regarding the life of John Haley Bellamy found within Ms. Smith's publication, the author has elected to omit this unverified testimony from the main body of this text.
20. Yvonne Brault Smith, *John Haley Bellamy, Carver of Eagles*, Pg. 19
21. Records of Arrests for the City of Portsmouth, New Hampshire, multiple volumes dating from 1876 to 1914, collection of the Portsmouth Athenaeum
22. *Navy Yard Orders, Form 5, Trade Card, Schedule "B," requisition number 817–119, John H. Bellamy*. National Personnel Records Center, Civilian Personnel Records
23. Moses Victor Safford, *The Magazine, Antiques*, Pg. 103
24. Personal correspondence from M. H. Neale, agency director, Maine Branch Office, New York Life Insurance Company, to John Haley Bellamy, September 24, 1907, collection of Dyer Library, Saco, ME
25. http://www.phoenixmasonry.org/masonicmuseum/fraternalism/maccabees.htm
26. http://www.phoenixmasonry.org/masonicmuseum/fraternalism/maccabees.htm
27. http.//www.nationmaster.com/encyclopedia/Washingtonian-movement
28. Horace Mitchell, "Eagles with Feelings," Pgs. 28, 30
29. Docket #54609: *Testimony of E.B. Eastman, M.D., regarding the mental incapacity of John H. Bellamy of Kittery, in the County of York*. Probate Court Records, York County, ME. In her book *John Haley Bellamy, Carver of Eagles*, author Yvonne Brault Smith takes liberties with this document and inexplicably elects to include the words "by reason of infirmity and mental incapacity" in her quotation of it. [See Yvonne Brault Smith, *John Haley Bellamy, Carver of Eagles*, Pg. 58] This phrase is nowhere to be found in this document.
30. Moses Victor Safford, *The Magazine, Antiques*, Pg. 105
31. Yvonne Brault Smith, *John Haley Bellamy, Carver of Eagles*, Pgs. 58–59
32. Yvonne Brault Smith, *John Haley Bellamy, Carver of Eagles*, Pg. 59
33. Receipt from S.[?] Pickering to John Haley Bellamy, June 2, 1906, collection of Dyer Library, Saco, ME
34. Receipt from Frisbee Brothers Meats and Groceries to John Haley Bellamy, February 26, 1909, collection of Dyer Library, Saco, ME
35. Receipt from H. F. Colby Dairy to John Haley Bellamy, November 1, 1907, collection of Dyer Library, Saco, ME
36. Receipt for Post Office Box 1, Kittery Point Post Office, June 30, 1907, collection of Dyer Library, Saco, ME
37. Personal correspondence from George Savary Wasson to Moses Victor Safford, n.d., private collection
38. Docket #54609: *Guardian Account, Schedule B, The Account of John D. Carty Guardian of John H. Bellamy of Kittery, in the County of York*. Probate Court Records, York County, ME

39. Kingwood College Library of American Cultural History
40. *Portsmouth City Directory*, 1913
41. Yvonne Brault Smith, *John Haley Bellamy, Carver of Eagles*, Pg. 59.
42. Last inscription found upon John Haley Bellamy's nightslate, (Acc# 1942.0471.000001) collection of The Mariners Museum, Newport News, VA

Epilogue

1. *Portsmouth Herald*, April 6, 1914; *Portsmouth Times*, April 6, 1914. These documents correct the mistaken claim that John Haley Bellamy died on April 6, 1914. [See: Yvonne Brault Smith, *John Haley Bellamy, Carver of Eagles*, Pg. 59] Both obituaries refer to Bellamy as having died on a Sunday. As their date of publication, April 6, 1914, fell on a Monday, it becomes obvious that the obituaries are reporting an event from the previous day, Sunday, April 5, 1914, thus establishing this as the date of Bellamy's passing.
2. *Portsmouth Herald*, April 6, 1914
3. *Portsmouth Times*, April 6, 1914
4. State of New Hampshire Certificate of Death for John Bellamy, #1435621, File Date April 5, 1914
5. *Portsmouth Herald*, April 9, 1914
6. *Portsmouth Times*, April 8, 1914
7. Horace Mitchell, "Eagles with Feelings," Pg. 30
8. Joseph W. P. Frost in "Bellamy Banter: The Life and Work of John Haley Bellamy," Sandra Rux and Joyce Geary Volk (editors). *Piscataqua Decorative Arts Society*, Vol. III, 2007–2009 Lecture Series (2011). Pgs. 1-2
9. Joseph W. P. Frost in "Bellamy Banter: The Life and Work of John Haley Bellamy," Sandra Rux and Joyce Geary Volk (editors). *Piscataqua Decorative Arts Society*, Vol. III, 2007–2009 Lecture Series (2011). Pgs. 1-2
10. *The Heritage of Woodcarving: An Exhibition of Decorative Arts by John Haley Bellamy, Master Ship Carver and Seven Rockland, Maine, Carvers, July 18 through September 7, 1975*. The William A. Farnsworth Library and Art Museum, 1975, Rockland, ME, Pg. 6
11. Smith noted George Michael, editor of National Antiques Review, as having been a prime inspiration for her interest in Bellamy.
12. E-mail communication with Arthur Swanson, woodcarver, restorer, and noted producer of Bellamy reproductions, February 27, 2009
13. Interview with Michael Dow, architectural woodcarver, restorer, and noted producer of Bellamy reproductions, July 20, 2008
14. E-mail communication with Michael Dow, architectural woodcarver, restorer, and noted producer of Bellamy reproductions, February 17, 2009
15. *Portsmouth Herald*, March 2, 1996
16. E-mail communication with Vincent Harrild, restorer of antiques and fine art, and noted producer of Bellamy reproductions, October 26, 2012
17. E-mail communication with Vincent Harrild, October 26, 2012

Bibliography

Primary Sources

Unpublished

Account of John D. Carty, guardian of John H. Bellamy of Kittery, Docket #54609: Guardian Account, Schedule A. York County Probate Court Records, York County, ME.

Bellamy, Charles Gerrish. Untitled autobiographical notes. Charles G. Bellamy File, Dyer Library.

———. "Letters of Incorporation" January 20, 1863. Charles G. Bellamy File, Dyer Library.

Bellamy, John Haley. "Atropos." Bellamy Papers and Correspondences, Farnsworth Art Museum and Wyeth Center.

———. Business agreement with David A. Titcomb, July 13, 1868. John Haley Bellamy Papers (1836–1914), Phillips Library, Peabody Essex Museum.

———. Business agreement with David A. Titcomb, July 30, 1866. John Haley Bellamy Papers (1836–1914), Phillips Library, Peabody Essex Museum.

———. Business agreement with Henry C. Merriam, November 26, 1868. John Haley Bellamy Papers (1836–1914), Phillips Library, Peabody Essex Museum.

———. Business agreement with Eldridge Walcott and Jonas C. Young, May 21, 1870. John Haley Bellamy Papers (1836–1914), Phillips Library, Peabody Essex Museum.

———. Business agreement with Eldridge Walcott and Jonas C. Young, May 30, 1870. John Haley Bellamy Papers (1836–1914), Phillips Library, Peabody Essex Museum.

———. "Declaration of intent to submit a patent to the U.S. Patent Office. . . August 8, 1868." John Haley Bellamy Papers (1836–1914), Phillips Library, Peabody Essex Museum.

———. "Declaration of intent to submit a patent to the U.S. Patent Office. . . January, 1873." John Haley Bellamy Papers (1836–1914), Phillips Library, Peabody Essex Museum.

———. "Description of a hauling device invention." Bellamy Papers and Correspondences, Farnsworth Art Museum and Wyeth Center.

———. "Design for a School House 40 ft. by 30 ft. ...January 1860, Charles G. Bellamy, Architect" Joseph W. P. Frost collection; photocopy, courtesy Earle G. Shettleworth, Maine Historic Preservation Commission, Augusta, ME.

———. "Little Wife." Bellamy Papers and Correspondences, Farnsworth Art Museum and Wyeth Center.

———. "So." Kittery Historical and Naval Society.

———. "Tomorrow." Bellamy Papers and Correspondences, Farnsworth Art Museum and Wyeth Center.

———. Untitled, undated poems, Farnsworth Art Museum; Wyeth Center; The Mariners Museum, Newport News, VA.

———. "Yes." Kittery Historical and Naval Society.

"Explanation of Plan of Old Pepperrell House, Kittery Pt., ME." Charles G. Bellamy File, Dyer Library.

Extracts from the Daily Log Book, US Navy Yard, Portsmouth, New Hampshire, October 15, 1819–December 17, 1929. Compiled by Captain David F. Boyd, USN, United States Naval Prison, Portsmouth, NH, March 31, 1931. NARA, Waltham, Massachusetts.

Invoice from Houghton Mifflin & Co.'s Periodicals to John Haley Bellamy, December 6, 1906. Charles G. Bellamy File, Dyer Library.

Knight of the Maccabees Beneficiary Certificate for Sir Knight John H. Bellamy, May 20, 1880. Bellamy Papers and Correspondences, Farnsworth Art Museum and Wyeth Center.

Letters Sent to the Bureau of Construction and Repair, 1842–1911, Vol. 7, from Aug 11, 1875 to Dec 1, 1877. RG 181 Naval Districts and Shore Establishments, Portsmouth Navy Yard, NH. National Archives and Records Administration.

List of Bellamy Family Birth and Death Dates. Bellamy Papers and Correspondences, Farnsworth Art Museum and Wyeth Center.

List of books associated with the Bellamy family. Charles G. Bellamy File, Dyer Library.

Navy Yard Orders, Form 5, Trade Card, Schedule "B," requisition number 817–119, John H. Bellamy. National Personnel Records Center, Civilian Personnel Records.

"Obituary for John Haley Bellamy." Unidentified, undated typescript. Bellamy Papers and Correspondences. Farnsworth Art Museum and Wyeth Center.

Phoenix Insurance Company of Hartford, Connecticut, Policy Number 192, December 19, 1869. Charles G. Bellamy File, Dyer Library.

Receipt of ten dollars from Charles C. Bellamy for filling the Town's Quota under the last call of the President, March 27, 1865. Dyer Library.

Receipt of ten dollars from Elisha Bellamy for filling the Town's Quota under the last call of the President, March 27, 1865. Charles G. Bellamy File, Dyer Library.

Receipt from S.[?] Pickering to John Haley Bellamy, June 2, 1906. Charles G. Bellamy File, Dyer Library.

Receipt for Post Office Box 1, Kittery Point Post Office, June 30, 1907. Charles G. Bellamy File, Dyer Library.

Receipt from H. F. Colby, Dairy to John Haley Bellamy, November 1, 1907. Charles G. Bellamy File, Dyer Library.

Receipt from Frisbee Brothers Meats and Groceries to John Haley Bellamy, February 26, 1909. Charles G. Bellamy File, Dyer Library.

Records of Arrests for the City of Portsmouth, New Hampshire, 1876–1914, MS 033. Portsmouth Athenaeum.

Record of Arrests for the City of Portsmouth, New Hampshire, 1887–1892 MS 033. Portsmouth Athenaeum.

Record of Arrests for the City of Portsmouth, New Hampshire, 1901–1905. MS 033. Portsmouth Athenaeum

Register of Marriages for Kittery Point, Maine, 19th Century. Special Collections, Portsmouth Public Library.

R.G. Dun & Co. Collection, Massachusetts, Vol. 79, p. 233. Baker Library Historical Collections, Harvard Business School.

Safford, Moses Victor. *Bellamy Genealogy.* Charles G. Bellamy File, Dyer Library.

———. *The Eagle Figurehead of the USS* Lancaster. Typescript, Mariner's Museum, Newport News, VA.

———. "Explanation of Plan of Old Pepperrell House Kittery Pt. Maine." Charles G. Bellamy File, Dyer Library.

———. *John Haley Bellamy, The Woodcarver.* Typescript. The Mariner's Museum, Newport News, VA.

Edward Shaw, *Civic Architecture* (2nd ed., Boston: Marsh, Capen & Lyon, 1834) copy with Charles G. Bellamy mss inscription owned by Earle G. Shettleworth, Jr.

Shettleworth, Earle G., "Notes on John Kutts," Maine Historic Preservation Commission, Augusta, ME.

Titcomb, D. A. "Statement of half interest in the Masonic frame inventions of John Haley Bellamy . . . November 2, 1868." John Haley Bellamy Papers (1836–1914), Phillips Library, Peabody Essex Museum.

Published

Art Students' League of New York, *Announcements for the Season of 1901–1902*, NYC, 1901.

Catalogue of the Officers and Students of the New Hampton Literary and Biblical Institution at New Hampton, N.H., for the Academical Year 1857–8. Concord, NH: Jones & Cogswell, Printers, 1858.

Charlestown Directory, 1864 and 1868.

Constitution of the New Hampton Literary and Biblical Institution. Gordon-Nash Library.

Department of Commerce and Labor Bureau of Navigation, *Thirty-Sixth Annual List of Merchant Vessels of the United States*. Washington: Government Printing Office, 1904.

Fernald, Mark. *Life of Elder Mark Fernald:* Newburyport, MA: George Moore Payne and D. P. Pike, Philadelphia: Christian general book concern, 1852.

Garth's Auction Inc., Delaware, Ohio. Auction records for item number 230306041681, John Haley Bellamy eagle sold on November 29, 2008.

History of the New Hampton Institution. MSS, Gordon-Nash Library.

Portsmouth City Directory, 1864 and 1913.

Safford, Moses Victor. "John Haley Bellamy, The Woodcarver of Kittery Point," *The Magazine, Antiques*, Vol. XXVII, no. 3 (March 1935). Pgs. 102–106.

"Tourists, Attention," Bellamy family handbill advertisement. Bellamy Papers and Correspondences, Farnsworth Art Museum and Wyeth Center.

U.S. Census: Schedule 1: Free Inhabitants in Kittery in the County of York State of Maine, June 1, 1860.

U.S. Census, Twelfth Census of the United States, Schedule 1–Population, Supervisors District No. 9, Enumeration District No. 239, Kittery, Maine, June 18, 1900. Department of Commerce and Labor–Bureau of the Census, 1900.

U.S. Census, Thirteenth Census of the United States: 1910–Population, Supervisors District No. 1, Enumeration District No. 246,Kittery, Maine, April 20, 1910. Department of Commerce and Labor–Bureau of the Census, 1910.

Wasson, David A. "The Silent Pilots," *The Outlook*, Volume CIX, (January 27, 1915): 213.

Winslow, Richard E., III. *"Do Your Job!" An Illustrated Bicentennial History of the Portsmouth Naval Shipyard, 1800–2000.* Portsmouth, NH: The Portsmouth Marine Society (published for the Society by Peter E. Randall, Publisher), 2000.

Winslow, Richard E., III. "A Race of Shipbuilders," in *The Hanscoms of Eliot, Maine.* Portsmouth, NH: Portsmouth Marine Society Press, 2013, chapters 9, 12–14.

Newspapers

New York Times
Portsmouth Daily Chronicle
Portsmouth Herald
Portsmouth Journal
Portsmouth Times
The Harbor and Island Echo, Reunion issue Nov. 27, 1895

Secondary Sources

Unpublished

McBrien, Johanna. "The Furniture Craft Community of Portsmouth, New Hampshire, 1798–1837." Univ. of Delaware MA thesis, Spring, 1996, unpublished typescript. Portsmouth Athenaeum.

Books

Babson, John J. *History of the Town of Gloucester, Cape Ann*. Gloucester, MA: Procter Brothers, 1860. Reprint, Gloucester, MA: Peter Smith Publisher, Inc. 1972.

Bardwell, John D. *Images of America: Old Kittery*. Dover, NH: Arcadia Publishing, an imprint of the Chalford Publishing Corporation, 1995.

Black, Frederick R. and Edwin C. Bearss. *The Charlestown Navy Yard, 1842–1890*. Boston, MA: Boston National Historical Park, July, 1993.

Boston, Charlestown Navy Yard, 1800–1842, historic resource study, volume I of II. Boston, MA: Boston National Historical Park, October, 1984.

Brewington, M. V. *Shipcarvers of North America*. Barre, MA: Barre Publishing Company, 1962.

Brighton, Ray. *Frank Jones King of the Alemakers*. Hampton, NH: Peter E. Randall, Publisher, 1976.

Candee, Richard M. "The Architecture of Maine's Settlement, Vernacular Architecture to About 1720" in *Maine Forms of American Architecture* (Deborah Thompson, ed.) Camden, Maine: Downeast Magazine for ColbyMuseum of Art, 1976.

Christensen, Erwin O. *Early American Wood Carving*. Cleveland, OH, New York, NY: The World Publishing Company, 1952.

Heritage of Woodcarving: An Exhibition of Decorative Arts by John Haley Bellamy, Master Ship Carver and Seven Rockland, Maine, Carvers, July 18 through September 7, 1975, The, Rockland, ME: The William A. Farnsworth Library and Art Museum, 1975.

Isaacson, Philip M. *The American Eagle*. Boston, MA: New York Graphic Society, 1975.

Larkin, Jack. *The Reshaping of Everyday Life, 1790–1840*. New York, NY: Harper & Row Publishers, Inc., 1988.

Lipman, Jean, and Winchester, Alice. *The Flowering of American Folk Art: 1776–1876*. New York, NY: The Viking Press, in cooperation with the Whitney Museum of American Art, 1974.

Pinckney, Pauline A. *American Figureheads and Their Carvers*. New York, NY: W.W. Norton & Company, Inc., 1940.

Safford, Moses Victor. *Kittery's Historical Background,* Kittery, ME: Kittery Press, 1936.

Sessions, Ralph. *The Shipcarver's Art: Figureheads and Cigar-Store Indians in Nineteenth-Century America*. Princeton, NJ, and Oxford: Princeton University Press, 2005.

Smith, Yvonne Brault. *John Haley Bellamy, Carver of Eagles*. Portsmouth, NH: Portsmouth Marine Society, 1982. Reprint, Portsmouth, NH: Portsmouth Marine Society, 2002.

Stackpole, Everett S. *Old Kittery and Her Families*. Lewiston, ME: Press of Lewiston Journal Company, 1903.

U.S. Portsmouth Naval Shipyard, *Cradle of American Shipbuilding*. Portsmouth, NH: Portsmouth Naval Shipyard, 1978.

Articles, Catalogs, and Essays

"Bellamy Banter: The Life and Work of John Haley Bellamy," Rux, Sandra and Volk, Joyce Geary (editors). *Piscataqua Decorative Arts Society*, Vol. III, 2007–2009 Lecture Series (2011).

Bourgeault, Ronald. "John Haley Bellamy 1836–1914," catalog. Northeast Auctions Marine & China Trade: Portsmouth, NH, August 21, 2005. Pg. 178

Candee, Richard M. "The New Colonials: Restoration and Remodeling of Old Buildings Along the Piscataqua," in *A Noble and Dignified Stream: The Piscataqua Region in the Colonial Revival, 1860–1930*, Sarah Giffen and Kevin Murphy, eds. York, ME: Old York Historical Society, 1992. Pp. 33–48

———. "Sparhawk Hall," *A Noble and Dignified Stream: The Piscataqua Region in the Colonial Revival, 1860–1930*, Sarah Giffen and Kevin Murphy, eds., York, ME: Old York Historical Society, 1992. Pp. 52–55

Eby, Cecil D., Jr. "Down East with George Savary Wasson." *DownEast Magazine,* Vol. XIII, no. 7 (April 1967). Pgs. 43–60.

Eckstorm, Fannie Hardy. "George Savary Wasson, Artist and Writer, 1855–1932." *Essex Institute Historical Collections*, LXXIX (1943). Pgs. 47–59.

Fairbrother, Trevor J, "Painting in Boston, 1870–1930," in *The Bostonians: Painters of an Elegant Age, 1870–1930*. Boston, MA: Museum of Fine Arts, Boston/Northeastern University Press, 1986. Pgs. 31–92.

Foss, R.W. Gerald D. "Samuel M. Dockum, Cabinetmaker." St. John's Lodge No. 1., F. and A.M., Portsmouth, New Hampshire, Vertical File 0142, Portsmouth Athenaeum.

Hatch, Francis W. "'Wave Crest' Retires." *DownEast Magazine*, Vol. XV, no. 9 (June 1969).

Hurlin, William. "The Academic and Theological Institution, New Hampton, N.H.," *The Granite Monthly,* January, 1900.

Hill, Nola. "Bellamy's Greatest Eagle." The Magazine Antiques, Vol. 51 (April 1947). Pg. 259.

Isaacson, Philip M. "John H. Bellamy," in *Versatility–Yankee Style: The Cultural Diversity of Rev. Jonathan Fisher, John H. Bellamy and the Hardy Family,* exhibition catalog. William A. Farnsworth Library and Art Museum, Rockland, ME, 1977.

Keyes, Homer Eaton. "Bellamy's Style and Its Imitators," *The Magazine, Antiques*, Vol. XXVII, no. 3 (March 1935). Pgs. 106–107.

Masonic Symbols in American Decorative Arts, exhibition catalog. Scottish Rite Masonic Museum of Our National Heritage, Lexington, MA, 1976.

Mitchell, Horace and Mitchell, Jeanette. "Bellamy's Eagles," *American Mercury* (September 1959). Pgs. 144–147.

Mitchell, Horace. "Eagles with Feelings," *Yankee Magazine* (September 1965). Pgs. 28–29, 30, 33.

National Park Service, Division of Publications. *Charlestown Navy Yard: Boston National Historical Park, Massachusetts*. Produced by the Division of Publications, Washington, DC: US Department of the Interior, 1995.

Port, Jane L. "Boston's nineteenth-century ship carvers." *The Magazine, Antiques*, Vol. CLVIII, no. 5 (November 2000). Pgs. 754–761.

Randall, Peter. "Its Time for Antiques" [re: Ivah W. Spinney], *New Hampshire Profiles* (Sept. 1967). Pgs. 54–56.

Ward, Ellen MacDonald. "Only a Memory," *DownEast Magazine*, 39 (February 1993), Pgs. 53–54.

Web sites

history.navy.mil
mariner.org
medfordhistorical.org
nationalacademy.org
nationmaster.com
pbs.org
phoenixmasonry.org

Appendix: Documented Bellamy Eagle Slogans

"DIRIGO"

"Don't Give Up The Ship"

"Dum Vivimus Vivamus"

"E Pluribus Unum"

"F.P.V.F.A."

"God Is Our Refuge And Strength"

"Happy New Year!"

"Health to the Navey"

"In God We Trust"

"In Hoc Signo Vinces"

"L.H. Wilson"

"Long Live Dewey!"

"Merry Christmas!"

"Nil desperandum Christo duce"

"Passaconaway. Tribe No. 5. Manchester. N.H. – T.O.T.E."

"Remember the Maine!"

"RENIM."

"The Lord is my Shepherd"

"Welcome Stranger"

"Wentworth L[odge] K.P."

Repainted Banners

"Anti-Volsted"

"Faith, Hope, and Charity"

Index

A

Acanthus leaf newel post, ca. 1870-80, 132, *133*

Adams, Captain Edward H., *91*, 125-126, *126*

Adams, Cass, 23, *23*

American Centennial carved box, 1876, 142, *144-145*

Appledore House Hotel, *65*, 80, *81*, 82

Architecture de Palladio, 42

Art Student's League of New York, 17

Atlantic Marine Exchange, 146-147, *147*, 162

Atlantic Monthly, 150, 168

B

Bangor, Maine, 168

Barber, Charles, 42

Barber, William, 42

Barnabee, Henry Clay, 148

Beecher, Laban Smith, 24-25, 35, 37, *17*, *19*, *143*
- *Architectural American Eagle*, *122*
- career as a woodcarver, 17-18
- influence upon John Haley Bellamy's artistic vision, 18-20, 67,143
- *Liberty Pole Eagle*, 17, *17*
- *Pepperrell Mansion Eagle*, 18, *19*, *143*
- *Portsmouth Navy Yard Marine Barracks Eagle*, 18, *19*
- probable associations with Charles Gerrish Bellamy, 18, *19*, 24-25
- USS *Constitution* figurehead of President Andrew Jackson, 17-18

Bellamy, Catherine (half-sister), 27, 30

Bellamy, Charles Carroll (brother), 26, 31, 40, 165, 175
- death, 165

Bellamy, Charles Gerrish (father) *9*
- career as architect, woodworker, house and boat builder, 9-11, 18, 20
- death, 165
- financial insolvency, 67-68
- history and political career, 10
- influence upon John Haley Bellamy's artistic vision, 11, 20
- working with John Haley Bellamy, 51, 67, 93, 137-138, *137*

Bellamy eagle plaque
- characteristics of, 104, 106
- origins of, 100, 102

Bellamy Eagle, the
- characteristics of, 84-85, 90, *112*, 113, *114-115*, 115, 123, 141
- composition and manufacture of, 114, *114*
- early commissions for, 80-81, 86-87
- importance of in American art, 80
- mass production of, 89-92, *91*
- origins of, 79-80
- prevalence of misattributions and forgeries, 113

Bellamy, Elisha (brother)
- assistant to John Haley Bellamy, 31-32, 51, 53, 56, 67, 93
- death, 165

Bellamy Family coat of arms, ca. 1870, 131-132, *131*, *181*

Bellamy, Frances Keene (mother), 9-10, *9*, 31, 67, 68, 80, 168
- death, 165

Bellamy, Frances (sister), 75, 165

Bellamy, John Haley
- adolescence/early adult years, *12*, 23, 25, 30-31, *32*
- apprenticeship to Samuel M. Dockum, 20, 21, 23, *23*, 27, 35, 52, 71, 141, 175
- arguments that his art be classified as Fine Art, 129-130, 184-185
- artistic legacy, 176
- artistic reputation of among contemporaries, 99, 132, 139, 145
- artistic training, 16
- assistants in shop, 51, 53, 67, 92, 93, 128, 152
- associations with American literati, 151; with Frank Jones, 79, 100, 102
- birth, 7
- carver of ship's figureheads, 38
- cheated out of his life savings, 172-173
- childhood of, *7-8*, 8-9
- death, 165, 171-75
- decorative woodcarver for merchant vessels, 18
- earliest artistic influences and artwork, 11
- early education, 12
- employment at the Portsmouth Navy Yard, 38, 40, 62-63, 72, 73-74, 79
- employment at the Charlestown Navy Yard, 41, 68-69, 79
- employment of Kittery-area tourism to further sales, 80
- employment with the Portsmouth Navy Yard, 38, 40
- failed attempt at courtship, 13, 16, 23, 25
- family history, 10
- first business venture as a woodcarver, 33
- Guardian of, John D. Carthy, 172
- house carpenter, 132, 137
- inventor, 47-51, 64, 160
- importance of in American art, 76-77, 176
- intellectual interests of, 155, 168
- later years, 159-160, *159*, *164*, 171
- legend of his stopping woodcarving
- marketing strategy of, 86-87, 89
- Masonic, Odd Fellow, and Papal-inspired carvings, *54-57*, *59*
- member of the Knights of the Maccabees, 170;
- member of the Washingtonian Inebriates, 170-171, *171*
- myth of his alcoholism, 169-170, 172
- myth of having attended art school, 16-17
- myths concerning his career as an inventor, 47, 49

myths concerning the personality of, 73-74
New Hampton Literary and Biblical Institution, 25-27, 33
painter, 151
personality of, 1, 12-13, 32-33
poet, 2, 156-161,*158*, *157, 159,* 161, *168*, 173
prices for artwork at auction, 1
rate of pay while an employee of Portsmouth Navy Yard, 72
reality of his alcoholism, 168-169, 170-171
reality of his stopping woodcarving, 166, 167, 171
reasons for his enrollment within New Hampton Literary and Biblical Institution, 25-26
reasons for leaving the workshop of Laban S. Beecher, 24-25
rediscovery of, 113, 180
relationship to Laban Smith Beecher, 17-18, 24, 35, 37, 67
source of his lumber, 92
source of income in later years, 167-168
suspicions of his having been a Freemason, 21
use of industrial production methods in his art, 53, 92
workshop at Portsmouth Navy Yard, 62-63, *63*
workshop, Pepperrell counting house, 102, *103*, *104*, 176-177, 179
as a yachtsman, 151

Bellamy, Josephine (sister), 165
Bellamy, Laura (sister), 75, 165
Bellamy, Martha Ann (sister), 165
Bellamy, Sophronia (sister), 75, 165
Bellamy's Style and its imitators, 126, 128-129, *129*
Benson, John Prentiss, 85, *86*
Boardman, Archie, 169
Bonhomme Richard, 9
Bonnie Doon, 153, *153*
Booth, Edwin, 148
Boston Catalog of Sales of Condemned Material, 146
Boston, Massachusetts, 11, 16, 28, *28*, 29, 37-38, 88, *90*, 111, 115-117, 126, 146-147, 162
Brown, Walter, 132, 134
Budd, Edward, 28
Burns, Tom, 148
Bush, President George, Jr., 182
Bush, President George, Sr., 182

C

Cap'n Simeon's Store, 150, 151
Carlyle, Thomas, 155
Carty, Dr. J.D., 171
Carved Eagle with Two Draped Flags Under a Center Shield, ca. 1872-1890, *22*, 88-89, *88*
Carved Lion Cat Head, pine, 1859, 35-36, *36*
Carved miniature books from captured Spanish warship, 142, *143*
Carved plaque with motif of roses, ca. 1864, 142, *143*
Challenge, 29
Charlestown, Massachusetts, 52-53, 57, *59*, 67, 79
Charlestown Navy Yard, 41, *44*, *75*
post- Civil War demobilization, 4, 41
work conditions, post-Civil War, 45, 46, 68, 79
Chelsea Clock Company, 182, *186*
Churchill, R.W., 175
Cider Hill Woodworks, Inc. See Swanson, Arthur
City Hotel, Portsmouth, New Hampshire, 81, 100
Clemens, Samuel (Mark Twain), 1, 148, 149, 151, 153, 156, 157
Clifford, Nathan, 26
Cohen, Senator William, 182
Coleridge, Samuel L., 155
Cole, Thomas, 178
Copy of a Bellamy Eagle, 1889, *151*
Cosgrove, Alice, *96*
Cowles Art School, Boston, 16
Crawford, Marion, 156
Cross, Jeremy, 51
CSS *Alabama*, 40
Curtis, James O., 37

D

David Copperfield, 155
Demers, Donald
Dexter, Gordon, 146
Dickens, Charles, 144
Dockum, Samuel M., 20-23, 65, 175
career as a woodcarver and cabinetmaker, 20-21, *22*, 23
possible instructor in Masonic symbolism to John Haley Bellamy, 21, 51
Don't Give Up the Ship eagle for the schooner *Gracie D. Buchanan*, 1888, 139, *141*
Don't Give Up The Ship eagle, ca. 1872-1900, *90*
Dow, Michael, 180-182, *182-184*
Dum Vivimus Vivamus eagle, ca. 1872-1907, 181, *182*
Dun & Company, R.G., 59, 61
Dyer Library, Saco, Maine, 2, 33

E

"Eagle Figurehead of the USS *Lancaster*," 2, 4, 10, 12-15, 34, 40, 42, 47, 62-63, 69 72-74, 76, 99, 102, 111, 113, 148-49, 156, 160, 175
Eagle frame, ca. 1866-72, *60*, 61
Eagle plaque, ca. 1872-80, 106, *108*
East Boston, Massachusetts, 37-38
Eagle with blank banner, 1886, 128, *129*
Eastman, Doctor E.B., 171, 172
Eliot, Charles, 148, 170
E Pluribus Unum eagle plaque, ca. 1872-1900, 104, 106, *107*

F

"Faith, Hope, and Charity" eagle, ca. 1872-1900, 127, *127*
Faxon, Lois M., 49
Fernald, Mark, 8-9
Flanders, Benjamin W., 26
Flying Dragon, The, ca. 1842, *11*, 12, 142
Fogg, George G., 26
Fort McClary, Kittery, Maine, 40
Fowle, Isaac, 17
Franklin Pierce Veterans Firemen's Association, *65*, 94, *97*
French, Daniel Chester, 150
French and Italian Notebooks, The, 155
Frisbee, Frank J., 172
Frost, Joseph William Pepperrell, 2, 26, 176-178, 182-183, *184*

G

God Is Our Refuge And Strength eagle plaque, ca. 1872-1900, 104, *105*, 106
Governor Ichabod Goodwin Mansion, Portsmouth, New Hampshire, 132, 134, *134*
Gracie D. Buchanan, 139, *141*
Grand Army of the Republic, 61, 123
Great Republic, 29
Green Shay, The, 150
Grimson, Robert, 183

H

H. C. Hopkins & Company Dry Goods and Furnishings Building, 86
H. Hall, Samuel, 37
Hamlin, The Honorable Hamlin, 148, 170
Hanscom, Isaiah, 4-5
Harrild, Vincent, 183, *186*
Hawk-on-a-Spar, ca. 1868, 134, 135-136, *135-136*
Hawthorne, Nathaniel, 155

Head of a "Fox"/ Canine Bust, ca. 1860, 130, *130*
Hichborn, US Naval Constructor, 68, 175
Home from Sea, 150
Homer, Winslow, 1, 148, 149, 170, 178
House of the Seven Gables, The, 155
Howells, William Dean, 1, 148, 149, 170

I
In God We Trust eagle plaque, ca. 1872-1900, 104, 106, *107*
In Hoc Signo Vinces eagle plaque, ca. 1872-1900, 106, *108*
Initial schematic drawings for USS *Lancaster* figurehead, ca. 1876-77, *15*
Initial schematic drawing for USS *Wabash* sternboard, ca. 1865-71, *45*
Isaacson, Phillip, 177

J
Jackson, President Andrew, 17-18
James, Henry, 151
Jefferson, President Thomas, 42
Jewett, Mary, 151
Jewett, Sarah Orne, 151
J.H. Bellamy & Co., 59, 61, 79
John Haley Bellamy, Carver of Eagles, 2, 178
"John Haley Bellamy, The Woodcarver of Kittery Point," 1-2, 7, 8, 10, 11, 12, 14, 16, 26, 30, 33, 40, 47, 62-63, 67, 69, 70-71, 73, 74, 76, 77, 93, 102-103, 110, 147, 155, 156, 167, 168
Jones, Frank, 79, 100, *100*, 102
Jones, John Paul, 8
Josephine, 21, 71

K
Kaselauskas, David, 182-183, *184-185*
Keyes, Homer Eaton, 113, 176
Kimball & Webster, 11
Kimball, Solomon, 169
Kittery Historical and Naval Society, Kittery, Maine, 178
Kittery, Maine
  arrival of tourism, 80, 160-161
  early history of, 1, 7
  as a summer resort for American literati, 148-153, 155-156
Kittery Point Bridge, 11
Kittery Point Free Will Baptist Church, 175, *178*
Kittery Point, Kittery, Maine, 87, 179, 182
  history of, 7-8
Knight, Richard, 28
Knights of Columbus, 61
Knights of the Maccabees, the, 170
Knights of Pythias, *64-65*
Kurtze [Kutts], John, 11

L
Lane, Fitz Henry, *28*, 29
Large American eagle atop an out-of-round ball, ca. 1872, *23*, *83*, *84*, *85*, 86, *86*, 136
Latrobe, Benjamin Henry, 42
Lawrence, Captain John, 166, 167, 171
Lewis, Frederick, 37, *38*
Life of Frederick Schiller, 155
Life of John Sterling, 155
Long Live Dewey! eagle, ca. 1898, 93, *94*

M
McCarthy, Paul, 118, *120*
McIntire, Samuel, 42
McKay, Donald, 29, *29*, 37
Mahogany eagle, ca. 1872-1900, 99, *99*
Maine State Seal. ca. 1872-1900, 132, *132*
Manchester, New Hampshire, 86, 88
Manson, Charles F., 169
Manson, Wayne, 178
Mariner's Museum, Newport News, Virginia, The, *77*, 162-163, *163*, 176
Markham, Edward, 151
Mary Willey, 139, *140*
Masonic frame, ca. 1866-72, *50*
Masonic frame, ca. 1870, *54*
Masonic frame with eagle, ca. 1866-72
Masonic Royal Arch Shelf Clock, ca. 1870, 53, *55*
Masonic shelf clock, ca. 1870, 53, *55*
Masonic shelf clock with Knights Templar symbols, ca. 1867, 53, *55*
Masonic What-Not shelf with Knights Templar symbols, ca. 1866-72, 141, *142*
Medford, Massachusetts, 37-38, 40, 61, 68, 150
Merriam, Henry C., 58
Merry Christmas eagle, ca. 1872-1900, 94, *95*
Mitchell, Horace, 113
Moose head, ca. 1868, 134, 135, *136*
Morgan, George T., 42
Morgan, William, 50
Morrison, John, 183, *185*
Mudge, Harriot, 172-173, 175

N
Nash, Stephen Gordon, 26
National Academy of Design, 17
New Hampton Literary and Biblical Institution, *24*, 26-27, 33
  history of, 25-26
New York City, New York, 17, 80
New York Life Insurance Company, 170
Nickerson, H.W., 175
Night Slate, n.d., *48*
Normal Art School, Boston, 16

O
Office sign with eagle, ca. 1889, *151*
Outdoor eagle executed in the round, ca. 1872, *82*, 86

P
Paint and Clay Club, 150
Palladio, Andrea, 42
Papal Clock, 61
Passaconaway Tribe No. 5. Manchester, N.H. eagle plaque, 1881, 106, 108, *109*
Peabody Essex Museum, Salem, Massachusetts, 2, 178
Pepperrell Arms, circa 1897, *155*
Pepperrell Hotel, the, 80
Pepperrell Mansion
  daily life for the Bellamy family while in residence, 8-9, *8*
  tourist attraction, 160-161, *160*
Pepperrell Mansion eagle, ca. 1830, 18, *19*, *143*
Pepperrell, Sir William, 94, 132, 160
Perry, Bliss, 151
Pilothouse eagle for the tugboat *Mitchell Davis*, ca. 1890, 141, *141-142*
Pine laurel wreath, ca. 1875, 132, 134, *134*
Plaisted, The Honorable H. Merrill, 26
Poems by John Haley Bellamy
  "Atropos," 161, 168
  "Bury Me Beneath the Rock," 158
  "Cenotaph," 158, *159*
  "Grandpa," 159
  "Little Wife," 157, *157*
  "So," 157-158, 161
  "To-Morrow," 158
  "Yes," 165-166, 168
Polk, President James K., 10
Portland, Maine, 86
Portsmouth Daily Chronicle, 99
Portsmouth Herald, 20, 35, 37, 68
Portsmouth Navy Yard, *66*

Portsmouth Navy Yard Marine
Barracks Eagle, 18, *19*
Portsmouth, New Hampshire, 7, 87-88, *87*, 169
Portsmouth, New Hampshire Liberty Pole Eagle, 17, *17*, 18
Portsmouth Times, 62
Pridham, John R, 119-120
Pring, Martin, 7
Pythagoras Lodge No. 33, Portsmouth, New Hampshire, 21, 51

Q
Quarterboard for the schooner *Golden Gate*, ca. 1869-1900, 140
Quarterboard for the schooner *Mary Willey*, ca. 1869, *140*

R
Reich, John, 42
Reina Mercedes, 142, *143*
Remember the Maine! eagle, ca. 1898, *93*, 94, 182
Roosevelt, Theodore, Assistant Secretary of the Navy, 146
Rowe, Captain Francis A., 4
Rusticators at the Cove, 151

S
Safford, Moses Victor
biographer of John Haley Bellamy, 1-2, 13, 23, 26, 47, 171, 175
see "John Haley Bellamy, The Woodcarver of Kittery Point," and "The Eagle Figurehead of the USS *Lancaster*"
St. Botolph Club, 150
Sargent, John Singer, 150
School of the Museum of Fine Arts, 16
Shelf Clock with Independent Order of Odd Fellows symbols, ca. 1870, *59*
Sir William Pepperrell/West Indies Wharf, 8, *103*
Smith, Captain John, 7
Smith, Earl, 162
Smith, Yvonne Brault, 2, 128, 178
Sparhawk Hall, 132, 134, *135*, *137-138*, 137-138
Sparhawk, Nathaniel, 132
Speedwell, 28
Spinney, Ivah W., 123, *124-125*, 125
St. Gaudens, Augustus, 150
Stiegel, H.W., 49
Swanson, Arthur, 121, 179-180, *181*

T
Taylor, Mary, 169
The Lord is my Shepard eagle, ca. 1872-1900, 94, *96*
Titcomb & Bellamy, 53, *54–56*, *59*
Titcomb, David A., 58-59, 67
Tokens & Treasures: Gifts to Twelve Presidents, 182, *184*
True Masonic Chart, 51
Twain, Mark. See Clemens, Samuel (Mark Twain)

U
United States Navy, 4-5, 18, 41, 72-73, 106, 142
USS *Alabama,* 18
USS *Concord,* 18
USS *Constitution,* 94
USS *Cumberland,* 40
USS *Enterprise,* 69, *70*
USS *Enterprise* gangway board, 1874, 69-70, *71-72*
USS *Essex,* 69
USS *Kearsage,* 40
USS *Lancaster,* 4-5, *77*, 110-111, *111*
USS *Lancaster* Eagle figurehead, 1879-1881, *77-78*, *163*
construction, 5, 28, 62-64
debut, 76
evolution and early design. 14-15, *15*
importance in American art, 76-77
initial schematic drawings, ca. 1876-77, *15*
USS *Ranger*, 8
USS *Santee,* 18
USS *Vandalia*, 72-73, *73*
USS *Wabash*, *44*, 46-47, 69
initial schematic drawings for sternboard, ca. 1865-71, *45*

W
Walcott, Eldridge, 58, 59
War & Peace Eagles, ca. 1872, 88-89, *88*
Ward, Christopher, 183
Ware, Isaac, 61
Washingtonian Inebriates, The, 170-171, *171*
Washington, President George, 42
Wasson, Amelia, 150
Wasson family coat of arms for the yacht, 153, *153*
Wasson, George Savary
artistic legacy, 182-183
author, 150-151
childhood, 149-150
critical praise for writings, 151
painter, 182-183
possible employee of John Haley Bellamy, 152
student of John Haley Bellamy, 151-152
Whaleback Light, ca. 1889, *156*
Webb, William, 29
Wentworth, "Long John," 26
Wentworth Lodge, Knights of Pythias. *65*
What-Not shelf, ca. 1868-1900, 141, *142*
Wiles, Mary Safford, 132
William A. Farnsworth Library and Art Museum, Rockland, Maine, 2, 177
Williams, Captain Zachariah, 139, *140*, 165
Williams, Eugene, 2, 8, 165
Williams, John, 119
"Willowbank," Kittery summer home of John Prentiss Benson, 65, 85, *86*
Wilson, George, 148
Wilson, T.D., US Naval Constructor, 14-15, 28, 63, 68
Wooden pen in the shape of a quill, ca. 1875, 142, *143*

Y
Young, Jonas C., 58, 59

About the Author

An independent curator and lecturer specializing in American marine art, author James A. Craig became enamored with New England's maritime saga while studying Anthropology at the University of Massachusetts.

Having by his own admission "fallen" into the world of fine art, Craig was first employed at the House of the Seven Gables Museum in nearby Salem, Massachusetts, where he served as a curatorial assistant and museum educator. He went on to serve as Associate Curator for Collections at the Cape Ann Museum in Gloucester, Massachusetts, where he published an award winning biography on the great American marine painter Fitz Henry Lane.

Besides *Fitz H. Lane: An Artist's Voyage through Nineteenth-Century America* (The History Press, 2006) he has published *100 Essex Road: an American Treasure* about the George Cabot Lee estate in Newton, Massachusetts (2008) and *Frank Vining Smith: Maritime Painting in the 20th Century* (Hard Press Editions, 2010).

In 2011 he contributed to the Portsmouth Historical Society's *Maritime Portsmouth: The Sawtelle Collection*, describing the artistic relationship of George S. Wasson and John Haley Bellamy based on his research for *American Eagle: The Bold Art and Brash Life of John Haley Bellamy*.

When not working as an independent consultant to fine art collectors, antique dealers, and museums, he can be found along the granite shores of the ancient fishing port of Gloucester, Massachusetts, where he resides with his wife Katie, their two cats, and that irascible scoundrel, George the Turtle.

Portsmouth Marine Society Books

1. *John Haley Bellamy: Carver of Eagles* by Yvonne Brault Smith
2. *The Prescott Story* by Ray Brighton
3. *The Piscataqua Gundalow: Workhorse for a Tidal Basin Empire* by Richard E. Winslow III
4. *The Checkered Career of Tobias Lear* by Ray Brighton
5. *Clippers of the Port of Portsmouth* by Ray Brighton
6. *"Portsmouth-Built" Submarines of the Portsmouth Naval Shipyard* by Richard E. Winslow III
7. *Atlantic Heights* by Richard M. Candee
8. *There Are No Victors Here: A Local Perspective on the Treaty of Portsmouth* by Peter Randall
9. *The Diary of the Portsmouth, Kittery, and York Electric Railroad* by John D. Bardwell
10. *Port of Portsmouth Ships and the Cotton Trade* by Ray Brighton
11. *Port of Dover: Two Centuries of Shipping on the Cochecho* by Robert A. Whitehouse & Cathleen C. Beaudoin
12. *"Wealth & Honour": Portsmouth During the Golden Age of Privateering* by Richard E. Winslow III
13. *The Sarah Mildred Long Bridge* by Woodard D. Openo
14. *The Isles of Shoals: A Visual History* by John D. Bardwell
15. *Tall Ships of the Piscataqua, 1830–1877* by Ray Brighton
16. *George Washington in New Hampshire* by Elwin W. Page
17. *Montgomery and the Portsmouth* by Fred Blackburn Rogers
18. *Tugboats on the Piscataqua* by Woodard D. Openo
19. *Rambles About Portsmouth* by Ray Brighton
20. *John Paul Jones and the Ranger* edited by Joseph G. Sawtelle
21. *"Constructing Munitions of War": The Portsmouth Navy Yard Confronts the Confederacy, 1861–1865* by Fred Blackburn Rogers
22. *In Female Worth and Elegance* by John F. LaBranche and Rita F. Conant
23. *Gosport Remembered: The Last Village at the Isles of Shoals* Edited by Peter E. Randall and Maryellen Burke, Ph.D.
24. *Heavy Weather and Hard Luck: Portsmouth Goes Whaling* by Kenneth R. Martin
25. *U.S.S. Albacore: Forerunner of the Future* by Robert Largess & James Mandelblatt
26. *"Do Your Job!" An Illustrated Bicentennial History of the Portsmouth Naval Shipyard, 1800–2000* by Richard E. Winslow III
27. *Jenny Lind and the Clipper Nightingale Figurehead* by Karl-Eric Svärdskog
28. *Portsmouth's Heyday in Shipbuilding* by Gertrude M. Pickett
29. *Just Rye Harbor: An Appreciation and History* by Thomas and Rosemary Clarie / edited by Peter Randall
30. *Friendly Edifices: Piscataqua Lighthouses and other Aids to Navigation 1771–1939* by Jane Molloy Porter
31. *Maritime Portsmouth: The Sawtelle Collection* edited by Richard M. Candee, appreciations by J. Dennis Robinson
32. *Under the Isles of Shoals: Archaeology and Discovery on Smuttynose Island* by J. Dennis Robinson
33. *"A Race of Shipbuilders": The Hanscoms of Eliot, Maine* by Richard E. Winslow III